Paleontological Resource Survey
Amistad National Recreation Area

Natural Resource Report NPS/NRPC/NRR—2009/133
(Public Version, September 2009)

Christy C. Visaggi
Dept. of Biology and Marine Biology
University of North Carolina Wilmington
601 S. College Road
Wilmington, NC 28403

Jack G. Johnson
National Park Service
Amistad National Recreation Area
4121 Veterans Blvd.
Del Rio, TX 78840

Angel S. Johnson
The Shumla School
117 Sanderson St.
PO Box 627
Comstock, TX 78837

Jason P. Kenworthy
National Park Service
Geologic Resources Division
PO Box 25287
Denver, CO 80225-0287

Vincent L. Santucci
National Park Service
Geologic Resources Division
PO Box 25287
Denver, CO 80225-0287

> **NOTE:**
>
> The Amistad National Recreation Area Paleontology Survey includes information regarding the scope and significance of the fossil record within Amistad National Recreation Area as well as paleontological resource management recommendations. Specific locality information is not included within this report and it can be distributed in the public domain.
>
> Paleontological resources are non-renewable (irreplaceable) pieces of the history of life on the Earth. Help the National Park Service preserve the sense of discovery for future generations:
>
> **Do not collect fossils, or any other item, within any unit of the National Park Service.** If you see a fossil, leave it in place, record the location, and share your discovery with a park ranger.

September 2009

U.S. Department of the Interior
National Park Service
Natural Resource Program Center
Fort Collins, Colorado

The National Park Service, Natural Resource Program Center publishes a range of reports that address natural resource topics of interest and applicability to a broad audience in the National Park Service and others in natural resource management, including scientists, conservation and environmental constituencies, and the public.

The Natural Resource Report Series is used to disseminate high-priority, current natural resource management information with managerial application. The series targets a general, diverse audience, and may contain NPS policy considerations or address sensitive issues of management applicability.

All manuscripts in the series receive the appropriate level of peer review to ensure that the information is scientifically credible, technically accurate, appropriately written for the intended audience, and designed and published in a professional manner. This report received formal peer review by subject-matter experts who were not directly involved in the collection, analysis, or reporting of the data, and whose background and expertise put them on par technically and scientifically with the authors of the information.

Views, statements, findings, conclusions, recommendations, and data in this report are those of the author(s) and do not necessarily reflect views and policies of the National Park Service, U.S. Department of the Interior. Mention of trade names or commercial products does not constitute endorsement or recommendation for use by the National Park Service.

Printed copies of reports in these series may be produced in a limited quantity and they are only available as long as the supply lasts. This report is also available from the NPS Natural Resource Publications Management site (http://www.nature.nps.gov/publications/NRPM/).

Please cite this publication as:

NPS 621/100191, September 2009

Contents

Appendices

Figures

Tables

Acknowledgements

We are extremely grateful to R. Slade, J. Labadie, and G. Garetz for their continued support of this project. J. Little and seasonal volunteers are appreciated for their assistance in the field. AMIS staff including L. Evans, F. Webster, S. Garard, and M. Webster provided additional support in completing this paleontological survey. Field assistance and access to fossiliferous outcrops offered by P. Dering, M. Harrington, and other members of the Shumla School are greatly appreciated. K. KellerLynn aided in obtaining new information on park paleontological resources as a result of her involvement in the Geologic Resource Inventory scoping meeting. C. Visaggi is indebted to S. Kline for his endless encouragement, her wonderful co-authors for their confidence in her work and guidance throughout this process, and C. Healy at the Student Conservation Association for making this happen. G. Bell and R. Slade graciously agreed to review this lengthy manuscript; we are much obliged for their comments and contributions to this report.

Dedication

Dedicated to Robert M. Linsley (1930-2006).

Thanks for your support, company, and conversation.
I will always cherish our love of snails.

~ Christy C. Visaggi

Photo courtesy of Colgate University.

Abstract

Amistad National Recreation Area (AMIS) is one of at least 219 National Park Service units with paleontological resources. Despite a rich and abundant fossil record comprising vertebrates, invertebrates, plants, microfossils, and trace fossils, paleontological resources remain relatively unexplored at AMIS. The purpose of this survey was to obtain adequate baseline paleontological resource data for the park and offer recommendations for the future management of AMIS paleontological resources.

This report was compiled through extensive fieldwork, literature review, and interviews with park staff. Fieldwork completed primarily in March 2006 covered 25 in-park localities and 11 additional exposures in the AMIS vicinity. The investigation focused mainly on Cretaceous deposits as limited records exist for these fossils on National Park Service land. Numerous publications detail Quaternary fauna and flora uncovered during archeological salvage operations prior to the construction of Amistad Dam.

The Mesozoic Era is represented by Lower and Upper Cretaceous units inside the park including the Salmon Peak Limestone, Devils River Limestone, Del Rio Clay, Buda Limestone, and Boquillas Formation. The most common marine invertebrates encompass rudistid bivalves, *Ilymatogyra* oysters, *Nerinea* gastropods, and *Cribratina* forams. Fragmentary echinoids, ammonoids, scallops, bryozoans, burrows, and borings are also found in these deposits. The most interesting fossils at the park are likely rudistid reefs, quite distinct from modern reefs which are composed primarily of coral, as well as ammonoids, as both of these organisms went extinct at the end of the Mesozoic.

The Cenozoic Era is represented by assemblages preserved in rock shelters formerly utilized by indigenous people of the Chihuahuan Desert. Plant remains, pollen, and vertebrates are commonly preserved in this cultural context; mollusks, arthropods, and middens are additionally noted in the literature at select localities. Paleontological resources of the Quaternary are often utilized in studying the diet and behavior of ancient cultures as well as changes in the biota and climate of southwest Texas.

There are several repositories for AMIS paleontological collections. Limited fossils recovered inside park boundaries as a result of this investigation were accessioned at AMIS headquarters; records for pre-existing fossils housed at the park were updated. Although current interpretative efforts mainly focus on cultural resources, opportunities for increased education and awareness of paleontological resources at the park are plentiful. Developing programs in cooperation with the neighboring Shumla School is suggested as a shared interest exists in learning more about the paleontology of the region. Heightened protection of *in situ* paleontological resources at the park is needed, particularly for six localities to be registered as significant with the NPS. Additional surveys set up through an inventory and monitoring program are highly recommended for AMIS as well. Threats such as theft, erosion, and development must be mitigated as these resources are non-renewable and are important for scientific research. Documentation of fossil localities, inventory of AMIS paleontological resources, improved fossil collections, and various recommendations resulting from this project impart a strong basis for the future of paleontology at Amistad National Recreation Area.

Introduction

Located on the border between the United States and Mexico in southwestern Texas (Fig. 1), Amistad National Recreation Area (AMIS) contains a rich paleontological record dating back 100 million years. Nearly all of the land encompassed by the park is composed of fossiliferous clays and limestones (Fig. 2A). The abundance of fossils inside the park has long been recognized; however, little information existed regarding these resources. The need to increase paleontological awareness encouraged AMIS to fund this park-specific comprehensive survey. Research objectives included a review of park geology, documentation of paleontological localities, assistance in collection updates, and recommendations regarding interpretation and management. This report is the first formal survey of paleontological resources at AMIS.

Paleontological resources are non-renewable (irreplaceable) pieces of the history of life on the Earth. Help the National Park Service preserve fossils for future generations: Do not collect fossils, or any other item, within any unit of the National Park Service. If you find a fossil, leave it in place, note the location, and share your discovery with a ranger.

Historical Background

Early inhabitants of the Pecos River region left their mark on the land of Val Verde County more than 10,000 years ago. Rock paintings, bison kills, hearth middens, and much more are discussed in countless archeological reports that document the hundreds of shelters in and around the AMIS vicinity (e.g., Anderson 1974). Spanish explorers arrived in the 1500s; Europeans did not attempt colonization until the 1800s. Most early settlements failed as a result of drought or frequent attacks by Native Americans. The mid-1800s brought about a military road and postal service to the remaining colonies in this region. The greatest development for southwestern Texas occurred in 1881 as the Southern Pacific Railroad bridged the gap between New Orleans and Los Angeles above the Pecos River (National Park Service 1974).

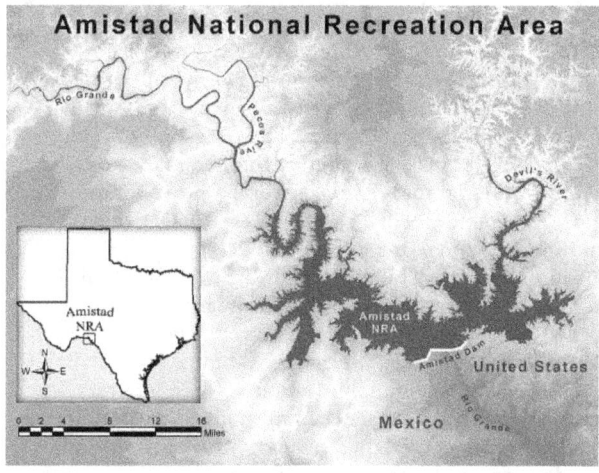

Figure 1. Location of Amistad National Recreation Area (Johnson and Johnson 2008).

The United States International Boundary and Water Commission (IBWC) and the Mexican government signed a joint treaty that affected the Rio Grande from El Paso to Brownsville in 1944 (Labadie et al. 2005). Proposed construction of hydroelectric dams came out of that agreement. Because important archeological localities might be inundated as a result of building these dams, archeological recovery operations commenced in 1958, led by the Texas Archeological Salvage Project (TASP). Archeological localities were given a Texas Archeological Site Number beginning with "41" for Texas and "VV" for Val Verde County, as is noted when referencing localities in this report. Many of the specimens cataloged in AMIS collections derived from excavations during this period prior to the completion of the dam in 1969. Publications from the late 1950s and early 1960s regarding resources collected as part of this salvage program refer to the lake as Diablo Reservoir. The name changed under the Eisenhower administration from "Diablo" (after the Devils River) to "Amistad" (meaning friendship in Spanish), serving as a more appropriate name for the international reservoir.

The park was established on November 11, 1965 as Amistad Recreation Area under a cooperative agreement with the International Boundary and Water Commission. It became authorized as a national recreation area on November 28, 1990. The elevation boundary of the park is 1,144.3 ft. regardless of normal pool level and most of the 23,674 ha (58,500 acres) managed by the National Park Service are underwater. The reservoir extends 40 km (25 miles) up the Devils River, 23 km (14 miles) up the Pecos River, and 117 km (73 miles) up the Rio Grande. AMIS receives 1.5 million visitors a year mainly for water-based recreation. Although most programs focus on the rock art and history of archeological excavations in the region, abundant paleontological resources surrounding the reservoir offer further opportunities for interpretation.

History of Paleontological Research

Marine fossils from southwestern Texas appeared in publications starting in the mid-late 1800s. Early reports commonly mention localities near Del Rio or along the Pecos River; thus it is likely that at least some research occurred inside boundaries presently defined by AMIS. Since the inception of the park in the late 1960s, paleontological resources from the Cretaceous Period found on National Park Service land were rarely reported in the literature. Fossils were not usually the focus of these few reports; paleontological resources are noted instead as part of stratigraphic sections.

One exposure of the Devils River Limestone measured shortly before the filling of the reservoir is now partially enclosed by AMIS (Smith and Brown 1983). Fossils from the overlying Del Rio Clay and Buda Limestone are additionally described in that publication; however, it is unlikely that they are exposed below the elevation boundary for the park. Smith and Brown (1983) further report on International Boundary and Water Commission (IBWC) cores of Salmon

Peak Limestone recovered near Amistad Dam that contain echinoids, brachiopods, clams, shell fragments, calcispheres (dinocysts), mixed forams, and abundant burrows as analyzed by C. H. Humphreys in the early 1980s. It is not clear if cores came from land presently managed by the National Park Service, but exploration into IBWC records could clear up this confusion. More on paleontological resources in the vicinity of AMIS can be found in similar geological road guides provided by Webster and Bolden (1983), Smith et al. (1984), Webster and Bolden (1984), and Spearing (1991).

Scientific publications by Coogan (1973), Humphreys (1984a, 1984b), Zahm et al. (1995a, 1995b), Kerans et al. (1995), Kerans (2002), Scott (2002a, 2002b), and Lock et al. (2007) signify research that may correspond to fossiliferous sections inside park boundaries. Although none of the authors mention the park, IBWC cores of Salmon Peak Limestone referenced earlier by Smith and Brown (1983) and exposures of the Devils River Limestone near the Pecos River may correlate to areas partially bounded by AMIS. Maps provided by Kerans (2002) strongly indicate sections enclosed by park boundaries. Fossils documented in these reports comprise echinoids, rudists, fragmentary mollusks, forams, and burrows created by crustaceans and worms. Thin sections utilized by Kerans and Scott are presently curated at UT Austin in the Texas Memorial Museum. Similar exposures described by Lock et al. (2007) may in part fall under NPS jurisdiction, but this cannot be confirmed based solely on maps in their publication.

The only report that specifically regards paleontological resources in Cretaceous deposits at AMIS apart from the present investigation is an NPS abstract by Jones (1993). He provides information on strata exposed inside AMIS and the fossils therein. It is not clear, however, if his descriptions are based exclusively on field research or if paleontological resources accounted for in these units are in part assumed from review of the literature. Fossils mentioned for the Lower Cretaceous Devils River Limestone and Salmon Peak Limestone include ammonites, rudists, clams, gastropods, echinoderms, corals, forams, ostracodes, and algae. Paleontological resources of the Upper Cretaceous Del Rio Clay comprise the characteristic oyster, *Ilymatogyra arietina*, clams, gastropods, forams, ostracodes, algae, and iron-stained trace fossils. The overlying Upper Cretaceous Buda Limestone and Boquillas Formation contain ammonites, bivalves, ophiuroids, echinoids, forams, algae, ostracodes, borings, and burrows. Rare fossils may include corals and bryozoans. The most recent reports of paleontological resources at AMIS are reviewed in an abstract by Visaggi (2006) prior to completing this publication and in an NPS Inventory and Monitoring report for the Chihuahuan Desert Network by Santucci et al. (2007).

Mesozoic paleontological resources are relatively unexplored at AMIS compared to specimens from the Pleistocene and Holocene epochs which were uncovered during salvage operations prior to the construction of Amistad Dam. Caves and bluff shelters developed in Cretaceous limestone, many of which also contain Quaternary fauna and flora, became endangered as plans for reservoir completion heightened in the 1950s. To avoid loss of archeological specimens due to inundation in and around the park, TASP initiated research and recovery of local resources as previously mentioned. Most specimens obtained as part of this rescue effort are now cataloged in AMIS collections held at the Texas Archeological Research Laboratory (TARL) in Austin. Although many of these specimens might be considered paleontological resources, most publications can be found in archeological journals. One exception is Lundelius (1984) as reviewed in the NPS report by Santucci et al. (2001) regarding paleontological resources found in caves.

Significance of Paleontological Resources

The most unique fossils in the park are undoubtedly rudistid reef-builders quite distinct from the coral reefs of our modern oceans. Rudists are an extinct group of anomalously large bivalves that existed exclusively during the Late Jurassic and Cretaceous periods of the Mesozoic Era (see Appendix D for a geologic time scale). The mass extinction at the Cretaceous-Paleogene (K-P) boundary may be renowned for the fall of the dinosaurs, but reefs composed of these odd-shaped bivalves also disappeared. Johnson (2001) recently questioned the role of rudists and scleractinians (corals) in Mesozoic shallow warm-water reefs based on sections of the Lower Cretaceous Devils River Limestone. Although his work did not incorporate fossils from AMIS, the abundance of these fossils in the park may prove fruitful in future research efforts. A wealth of information on rudistid bivalves can be found in recent publications by Filkorn et al. (2005) and Scott (2007).

Exposures of the Devils River Limestone in the AMIS vicinity exhibit a range of reef development among rudists from hydrodynamic accumulations to isolated mounds that can reach up to 5 m (15 ft) in height (Kerans 2002). Scott (1990) recorded four genera of rudists in rocks of the upper Albian Stage in North America (*Caprinuloidea*, *Mexicaprina*, *Texicaprina*, and *Kimbleia*). Type sections and specimens referenced by Scott (2002a) from the Pecos River valley may have originated in rock now under NPS jurisdiction; further investigation is needed to confirm precise locations. These peculiar bivalves are additionally noted in the Lower Cretaceous Salmon Peak Limestone at select areas in and around AMIS.

Ilymatogyra arietina oysters characterize the Upper Cretaceous Del Rio Clay. Although such oysters are frequently observed in the neighboring Grayson Formation of north-central Texas, specimens are significantly smaller than their southwestern relatives. The Grayson Formation is renowned for its "dwarfed" fauna exhibited in oysters, ammonites, and crinoids among several other taxonomic groups (e.g., Pampe 1979). Scott (1924) initially believed ammonite individuals might be stunted as a result of anaerobic conditions, but later regarded shells as small species or the preserved inner whorls of larger specimens (Scott 1940a). Kummel (1948) advocated that excessive amounts of iron might be the cause of stunted growth, but Mancini (1978b) disputed this notion as foraminiferal evidence suggests the presence of normal marine conditions instead. He commented that *Ilymatogyra* is paedomorphic in that it develops mature ornamentation at an earlier stage in life (Mancini 1977). The reduced size might be a modification for softer substrates as ubiquitous individuals from the in part equivalent Del Rio Clay in southwestern Texas are of normal size having perhaps lived upon firmer substrates in a nearshore environment (Mancini 1975, 1977). More on soft sediment adaptation of such fauna is described at length by Mancini (1978c). This substrate hypothesis is partially supported by Willis (1997) in McLennan County based on echinoderms of normal size, diminutive cephalopods, gastropods, scaphopods, and bivalves, as well as the presence of *Chondrites* ichnofossils (trace fossils of highly branched burrows). Although direct comparisons are not made to deposits of the Del Rio Clay in Val Verde County, research involving AMIS sections might offer new insight into this controversial subject.

The Upper Cretaceous Buda Limestone is highly fossiliferous in and around AMIS, yet reports of fossils from this unit are chiefly based on deposits from north-central Texas. Several publications comment on sections near Del Rio that are utilized in sequence stratigraphic research by Trevino (1988), Trevino and Smith (2002), and Lock et al. (2007). Buda Limestone exposures examined in this investigation contain an exceptionally diverse fauna of regular echinoids, irregular echinoids, ammonites, gastropods, bivalves, and trace fossils (Fig. 2C).

The Upper Cretaceous Boquillas Formation in Terrell County and Val Verde County contain stratigraphically important ammonites and inoceramid bivalves as documented by Freeman (1961). Sections near Del Rio examined by Trevino (1988) and Trevino and Smith (2002) include burrows, forams, calcispheres, echinoids, ammonites, and inoceramids. Revised paleoenvironmental interpretations for the area by Lock and Fife (2004), Peschier (2006), and Lock and Peschier (2006) incorporate additional fossils including saccocomid crinoids. Although elevated exposures of this unit are difficult to access within the park, locally important sections demonstrate the potential significance of fossils from the Upper Cretaceous Boquillas Formation in the vicinity of AMIS.

Fossils of the Quaternary Period of the Cenozoic Era found at Cueva Quebrada (41VV162A) are among the largest AMIS collections held at TARL. Located in a small canyon near the Pecos River, recovery of this terrestrial mammalian fauna occurred beyond park boundaries (Lundelius 1984). Specimens are unique in that most fragments are extremely burned; however, it is not clear if humans or spontaneous combustion of packrat middens are responsible for charred bones. The accumulation of material in the cave is similarly questionable as it may be due to carnivorous animals or anthropogenic activity. Although few fragments contain evidence of predatory feeding, lack of hearths negates human involvement (Lundelius 1984). The unusual occurrence of the Cueva Quebrada mammalian fauna remains unsolved.

Bonfire Shelter (41VV218) in Mile Canyon is another important source of ancient material near AMIS. Quaternary terrestrial specimens from this locality, however, are cataloged in park archeological collections. Fragments of hundreds of bison individuals from this shelter suggest that hunting here benefited from the "jump" method as humans drove bison herds off of the overhanging cliff into the shelter (Dibble and Lorrain 1968). Charred bone, projectile points,

and other cultural remnants indicate how people of the Pecos repeatedly utilized this cave for bison kills. Other land fauna and flora are commonly found at Bonfire Shelter; pollen grains are often utilized in paleoclimate reconstructions. These suggest the early presence of a pine woodland that eventually progressed into an increasingly arid Chihuahuan Desert (Bryant 1977). Although chiefly recognized for its archeological significance, specimens recovered from Bonfire Shelter could be considered in part paleontological resources.

Late Quaternary changes in vegetation are often revealed through palynological (pollen and spore) analyses (e.g., Bryant 1968; Bryant and Larson 1968). Pollen samples obtained from AMIS localities include Centipede Cave (41VV191) and Damp Cave (41VV189) (Johnson 1959, 1963). Changes in faunal composition observed at other shelters in the vicinity, such as Zopilote Cave (41VV216), Eagle Cave (41VV167), Coontail Spin (41VV82), and Devil's Mouth (41VV188), additionally suggest a shift to heightened aridity and erosion throughout the Quaternary (Raun and Eck 1967). Fruit, seeds, pollen, and small bones found in coprolites of Hinds Cave (41VV456) not only offer insight into early human diets, but speak of the fauna and flora that dominated this desert landscape (Williams-Dean 1978; Stock 1983; Reinhard et al. 2003; Dean 2006).

Other significant paleontological resources found in Quaternary settings include abundant fire-cracked rock (FCR). These colorful red rocks altered by thermal means are usually uncovered in archeological localities referred to as burned rock middens (BRM). Most FCR at the park originated from the very fossiliferous Del Rio Clay of the Upper Cretaceous; it seems that these mollusk-rich rocks were preferentially used in hearth features (Fig. 2B). The reason for this selection (as postulated by Dr. Don Lewis in 1995) might be that fossil-bearing rocks are less explosive at high temperatures compared to non-fossiliferous limestone due to differences in composition (Labadie 2004). Being as the majority of fire-cracked rock is a result of exposure to extreme heat conditions achieved in cooking, it certainly would have been advantageous for inhabitants of the Lower Pecos to have utilized the less explosive fossiliferous rock at the hearth instead. Over 700 localities contain fire-cracked rock at AMIS (Labadie 2004). Although archeological resources are the highlight of interpretive programs at AMIS, paleontological resources are a promising source of scientific and public interest at the park. Fossils from Cretaceous units inside the park offer potential for important discoveries as little research has commenced thus far at AMIS despite the prevalence of exposed fossiliferous sections. Quaternary resources primarily recovered during salvage operations at local rock shelters are however often well-documented. Extensive canyon sections provide a multitude of opportunities for evolutionary, paleoecological, and paleoenvironmental investigations. The fields of sequence stratigraphy and biostratigraphy are other appealing avenues of research particularly if ammonoids and microfossils can be utilized for zonation schemes. This is merely the beginning of unraveling the relatively unexplored fossil record at Amistad National Recreation Area; a bright and fossiliferous future lies ahead.

Figure 2. Sample of AMIS fossil resources. A) The senior author explores Salmon Peak Limestone along the Devils River. B) Fire-cracked rock full of *Ilymatogyra arietina* oysters found inside the park. C) Ammonite impression exposed in the Buda Limestone. Photos by Jack Johnson.

Figure 3. Stratigraphy of Cretaceous units exposed at Amistad National Recreation Area. After Kerans (2002) and Scott (2002b).

Geology

Fluctuating sea levels in the Cretaceous Period (144 - 65 million years ago) left a rich fossil record of marine and non-marine deposits stretching southwest to northeast across central Texas (Barnes 1992). Evidence of dinosaurs, mammals, reptiles, amphibians, birds, and an array of oceanic life is preserved in these rocks (Spearing 1991). Flowering plants (angiosperms) additionally flourished in warm and humid climates of the Cretaceous. Tectonic forces responsible for connecting the Gulf of Mexico and the Arctic Ocean through the Western Interior Seaway later resulted in uplift of the Rocky Mountains during the Laramide Orogeny (Spearing 1991). Trackways of land-dwelling fauna are often observed along the edge of this major ancient seaway that covered much of America's Heartland during the last 30 million years of the Mesozoic Era. The area now known as Trans-Pecos Texas lay at the connection between this ancient seaway and the ancestral Gulf of Mexico (Scotese 2001). During building of the Rocky Mountains shortly after the end of the Cretaceous, the entire region was lifted well above sea level and deposition of marine rocks ended.

Geologic History

Amistad National Recreation Area is underlain by rocks of the Lower (older) and Upper (younger) Cretaceous in the physiographic province called the Edwards Plateau. This region chiefly consists of fossiliferous shales and limestones bounded on the south and east by the Balcones Escarpment. Diversity of modern biota is partially attributed to the geography of the reservoir at the convergence of the Chihuahuan Desert, Edwards Plateau Savannah, and Tamaulipan Mezquital ecoregions. Marine fossils from the Cretaceous are entombed in bedrock of limestone canyons carved by rivers over millions of years after regional uplift; paleontological resources of the Quaternary are commonly preserved within these rock shelters formerly utilized by indigenous people. Ongoing erosion of these rocks exposes ancient fauna and flora from both the Mesozoic and Cenozoic, exhibiting the wealth of paleontological resources available at AMIS.

The grand cliffs that bound the park are composed primarily of Lower Cretaceous Devils River Limestone and Salmon Peak Limestone. These units represent shallow marine reef deposits in and around the Maverick Basin (Lozo and Smith 1964, Jones 1993). Outcrops of Devils River Limestone stretch in a band along the Rio Grande from western Texas to the Pecos River (Smith et al. 2000), upon which exist some of the most complete exposures of Lower Cretaceous rock in North America (Kerans et al. 1995). Facies change into the younger and in part laterally equivalent Salmon Peak Limestone occurs eastwardly from the Pecos River. This variably shelly lime mudstone encircles the main reservoir area and much of the Devils River (Barnes 1977). Spearing (1991) informally categorizes Lower Cretaceous rocks near Del Rio as the Santa Elena Limestone based on lithological similarities to canyons at Big Bend National Park; however, most early publications regard these same units as the Georgetown Limestone instead (e.g., Freeman 1964a, 1964b).

Lozo and Smith (1964) replaced this latter name in southwestern Texas, but rocks of this age in other parts of the state are still referred to as the Georgetown Limestone. Although older fossiliferous deposits including the McKnight Formation and West Nueces Formation are recorded in the AMIS vicinity based on IBWC cores (Smith and Brown 1983; Zahm et al. 1995a, 1995b), such deposits are not discussed further in this report as they are not exposed at the surface inside park boundaries.

Upper Cretaceous rocks unconformably overlie the Devils River Limestone and Salmon Peak Limestone at AMIS (Jones 1993). Lock et al. (2007) note that in Terrell and Val Verde counties, abundant *Gastrochaenolites* bivalve borings mark this unconformity. The basal unit exposed in the park of the Upper Cretaceous is the Del Rio Clay, a shallow nearshore deposit easily recognized by abundant *Ilymatogyra arietina*. This oyster, previously classified in the genus *Exogyra* by Boese (1919) and many others, is sometimes referred to as the "Devil's Toenail" due to its characteristic curled shell. Although the Del Rio Clay is mapped in a number of areas at AMIS (Barnes 1977), scattered rock fragments along the shoreline are more common than complete sections. Prior to the naming of the clay in southwestern Texas, stratigraphers classified this unit as the Grayson Formation of north-central Texas. The Grayson does not correlate unequivocally to the Del Rio Clay, however, as part of the overlying Buda Limestone is likely incorporated within the Grayson Formation (Stephenson 1944).

The Buda Limestone rests unconformably on the Del Rio Clay and represents a shift to deeper environments in southwestern Texas (Jones 1993). This fossiliferous unit as well as undivided Del Rio Clay and Buda Limestone are mapped throughout the park (Barnes 1977). The overlying unconformable Boquillas Formation exhibits continued movement of deposition offshore (Jones 1993; Lock and Fife 2004; Lock and Peschier 2006; Peschier 2006). Fissile shales are common in this distinctive flagstone exposed at higher cliff faces along the reservoir near Zuberbueler Bend (Barnes 1977). Small pockets of this unit may be found in other park areas (Barnes 1977), but outcrops are likely not accessible if at all exposed. Facies change from the Boquillas Formation into the in part laterally equivalent Eagle Ford Group occurs near the eastern edge of the reservoir offering little potential for exposures inside park boundaries. Both the Boquillas Formation and Eagle Ford Group are regarded as fossiliferous; however, few reports focus on paleontological resources in these deposits near AMIS. Finally, rocks of the Upper Cretaceous end in the AMIS vicinity with the Austin Chalk. Exposures of this unit are not likely found inside park boundaries; nonetheless, potential exists for local fossil discoveries (Barnes 1977).

The mass extinction at the end of the Cretaceous is not preserved in the sections at AMIS and neither is the recovery that followed during the Paleogene Period (Tertiary). Neogene Period (Tertiary) representation is likely limited to the caliche-cemented Uvalde Gravel, which is not known to be fossiliferous. Deposition of this unit occurred either during the Pliocene? and/or Pleistocene? Epoch of the Quaternary Period. Refer to Appendix D for a geologic time scale. Pleistocene fluviatile (river) terrace deposits and alluvium that accumulated during the Holocene Epoch are not usually regarded as fossil-bearing units; however, rock shelters carved in Cretaceous limestone commonly preserve an array of Quaternary fauna and flora in and around the park. These cultural deposits are often investigated by archeologists, yet specimens recovered from these localities can be utilized as paleontological resources (e.g. paleoclimate). Although the rock record may seem sparse for much of the Cenozoic, a long history of regional active geology and erosion is demonstrated by the magnificent and deeply incised canyons that presently surround Amistad Reservoir.

The following sections report on the stratigraphy of Cretaceous units exposed at AMIS (Fig. 3), a chronological list of confirmed in-park paleontological resources for the Cretaceous and for cultural deposits based on NPS excavations listed in publications (Table 1), and a review of fossils by formation as recorded nearby or elsewhere in Texas.

Fossiliferous Rock Formations
Fauna reported as part of this investigation are not included in the following descriptions. Park fossils are confirmed instead in the sections *Paleontological Resources Inventory* and *Localities*. A brief review of AMIS fossils is listed also in Table 1. Work by Jones (1993) is not described here either as his summaries did not correspond unequivocally to specific rock formations (as mentioned in the introductory chapter of this report). Fossils documented by Smith and Brown (1983) for Lower Cretaceous sections within the park are reviewed in the introduction, the following formation descriptions, and the inventory of paleontological resources for AMIS. Taxa listed in publications by authors such as Kerans (2002), however, are restricted to the introductory chapter and the following paragraphs as it is not yet clear if such fossils derived from sections inside the park.
Geologic formations exposed inside park boundaries are listed in Table 1 and Figure 3. Discussions of the Eagle Ford Group and Austin Chalk are restricted to the following section describing fossiliferous rock formations of the region as neither has been definitively identified within the park.

Devils River Limestone
(Lower Cretaceous: Early - Late Albian)
Lithology: Locally dolomitized limestone deposited as a margin reef along the rim of the Maverick Basin (Jones 1993). Mounds of rudists are prevalent in the upper part of the section; nodular limestone is common near the base (Barnes 1977). Changes in faunal composition represent a shift from mud- to grain-dominated environments (Kerans and Zahm 1998). Thickness is about 210 m (700 ft). (Barnes 1977).

Fossils: Diverse molluscan faunas include ammonites (Young 1961, 1979; Scott 2002b), oysters (Kerans and Loucks 2002), large chondrodont and rudistid bivalves (Miller 1984; Kerans and Zahm 1998; Kerans and Loucks 2002; Scott 2002b). Mollusks in and around AMIS are noted further by Smith and

Brown (1983) and Kerans (2002). Four genera of rudists described by Scott (1990) after Coogan (1973) are *Caprinuloidea*, *Kimbleia*, *Mexicaprina*, and *Texicaprina*. Other fossils reported nearby for this formation include corals (Kerans and Zahm 1998; Johnson 2001) and miliolid benthic forams (Kerans and Zahm 1998; Johnson 2001; Scott 2002b).

Salmon Peak Limestone
(Lower Cretaceous: Middle - Late Albian)
Lithology: Lower mudstone characterized by *Globigerina*; upper part cross-bedded, granular limestone, dominated by caprinid rudists and other mollusks (Barnes 1977). Unit is described by Hovorka (1996a, 1996b) as poorly cyclic and fairly homogenous; however, sequence stratigraphic analyses by others suggest these hemipelagic intrashelf basin deposits can be subdivided into five lithofacies instead (Zahm et al. 1995a, 1995b; Zahm 1997). Thickness about 95 m (310 ft). (Barnes 1977).

Fossils: Fauna reported in the vicinity of AMIS encompass caprinid rudists, mollusks, brachiopods, and echinoids (Barnes 1977; Smith and Brown 1983; Humphreys 1984a, 1984b; Zahm et al. 1995b). Microfossils (Smith and Brown 1983; Humphreys 1984a, 1984b) comprise planktonic forams *Globigerina*, *Hedbergella*, and *Favusella* (Barnes 1977; Zahm et al. 1995b), and benthic forams *Ammodiscus*, *Lagena*, *Haplophragmoides*, *Gavelinella*, *Lenticulina*, and *Spiroplectamina* (Zahm et al. 1995b). Calcispheres (dinocysts) are additionally noted (Smith and Brown 1983) and (Humphreys 1984a, 1984b). Ichnofossils (trace fossils) include *Skolithos*, *Planolites*, and *Thalassinoides* (Smith and Brown 1983; Humphreys 1984a, 1984b; Zahm et al. 1995b).

Del Rio Clay
(Upper Cretaceous: Early Cenomanian)
Lithology: Transgressive unit deposited in a shallow nearshore muddy environment (Jones 1993). Characterized by abundant *Ilymatogyra arietina* oysters; iron oxide minerals common. Some lenticular beds of highly calcareous siltstone may be present, but chiefly a gypsiferous or calcareous clay. Thins to the northwest. Thickness up to 61 m (200 ft). (Barnes 1977).

Fossils: *Ilymatogyra arietina* (Roemer) 1849 is the highly abundant and characteristic oyster of the Del Rio Clay (Offeman et al. 1982). Other fauna include ammonites (Kennedy et al. 2005), gastropods (Maddocks 1988), echinoids (Lock et al. 2007), decapods (Richardson 1955), and ostracodes (Maddocks 1988). Microfossils comprise forams (Loeblich and Tappan 1946; Bullard 1951; Maudlin 1985) and coccolithophores (Hill 1975). Trace fossils are also documented (Udden 1908) and *Gastrochaenolites* bivalve borings mark the base of the section (Lock et al. 2007). Fossils from the partially-equivalent Grayson Formation encompass forams, ostracodes, sponges, corals, bryozoans, bivalves, gastropods, nautiloids, ammonoids, belemnoids, asteroids, echinoids, crinoids, annelids, and an assortment of fish as summarized by Mancini (1978a, 1978b, 1982). Scaphopods and ichnofossils are further reported (Willis 1997). The renowned "dwarfed" organisms and controversies regarding paedomorphism or adaptations to the substrate are discussed by Scott (1924, 1940a), Kummel (1948), Mancini (1975, 1977, 1978b, 1978c), Pampe (1979), and Willis (1997). Fauna utilized for biostratigraphic and paleoecologic analyses in the Grayson are noted by Tappan (1939, 1940), Loeblich and Tappan (1946), Albritton et al. (1954), Mancini (1979), Aepli (1989), Willis (1997), and Cheetham et al. (2006). More on the paleontology of the Grayson Formation can be found in the AMIS chapter of the Chihuahuan Desert Network report by Santucci et al. (2007).

Buda Limestone
(Upper Cretaceous: Early Cenomanian)

Lithology: Fine-grained, bioclastic, commonly glauconitic or pyritiferous limestone. Hard, massive, and poorly bedded into nodular, argillaceous, and thinly bedded. Trace fossils and bivalves common. Thickens eastward. Thickness 14 – 30 m (45 - 100 ft) (Barnes 1977).

Fossils: Faunal assemblages consist of calcareous algae, sponges, corals, bryozoans, bivalves, gastropods, ammonites, echinoids, and thalassinoidean burrows (Stephenson 1944; Reaser and Dawson 1995a, 1995b; Reaser and Robinson 2003). Diverse cephalopods are noted by Archer (1936), Young (1962), Hook and Cobban (1983), and Kennedy et al. (2005). Stromatoporoids (Schmidt Murphy 1991), calcisponges (Wells 1934), *Microsolena* corals (Wells 1944), and borings made by bivalves such as *Gastrochaena* are additionally documented (Reaser and Dawson 1995b).

Boquillas Formation
(Upper Cretaceous: Middle Cenomanian - Turonian)

Lithology: Four subunits are described by Barnes (1977), all combinations of interbedded silty shale, siltstone, and clastic or granular limestone. Tepee structures are common and reflect an upward displacement of laterally confined rock that expands in response to surface caliche formation (Lock and Choh 1997; Lock et al. 2001). Deposit historically referred to as the Boquillas Flags. Thickness is between 49 and 67 m (160 - 220 ft) Facies gradually changes to that of the Eagle Ford Group moving east across the Devils River (Barnes 1977).

Fossils: Inoceramids, ammonites, echinoids, crinoids, calcispheres, planktonic forams, and trace fossils from Val Verde and Terrell counties are used in both stratigraphic and paleoenvironmental investigations (Freeman 1961; Trevino 1988; Trevino and Smith 2002; Lock and Fife 2004; Lock and Peschier 2006; Peschier 2006). Marine vertebrates (Cicimurri and Bell 1996; Bell 2002), cephalopods (Young 1958; Powell 1967; Hook and Cobban 1983; Young 1984; Cobban 1988; Kennedy and Cobban 1993), bivalves (Cobban and Hook 1980; Ashmore et al. 1997; Kennedy and Cobban 1993), as well as microfossils (Huffman 1960; Frush and Eicher 1975; Graham 1995, 1997) are additionally reported elsewhere in the Boquillas Formation.

Eagle Ford Group
(Upper Cretaceous: Cenomanian -Turonian)

Lithology: Flaggy siltstone, limestone, and fine-grained sandstone in lower section; flaggy shale and limestone in upper section. Laminations commonly observed in the basal part of the unit. Thins northeastward. Thickness 23 – 61 m (75 - 200 ft). (Barnes 1977).

Fossils: Microfossils are utilized in stratigraphic analyses by Moreman (1927, 1942), Bostik (1960), and Pessagno (1969). These assemblages may comprise pollen (Brown and Pierce 1962; Stone 1967; Christopher 1982), forams (Schell 1952; Jones 1960; Loeblich and Tappan 1961; McNulty 1964; Longoria 1973; Steinman 1974; Barrett and Goodson 2006; Nebrigic 2006) and ostracodes (Crane 1965; Hazel 1969). Mollusks including cephalopods (Moreman 1927; Scott 1940b; Moreman 1942; Stephenson 1955; Clark 1960; Powell 1963; Kennedy 1988), gastropods (Stephenson 1955), and bivalves (Moreman 1942; Stephenson 1955; Kauffman 1965; Cobban and Hook 1980; Friedman and Hunt 2004) are often recorded. Fossil pearls are discussed by Friedman and Hunt (2004). Other marine invertebrates include coral (Perkins 1951), crabs (Bishop 1989), and crustaceans (Vega et al. 2007). Vertebrates are reported by numerous authors including Bilelo (1969), Meyer (1988), and Friedman (2001b). Fish are most commonly documented (McNulty 1970; Gilette 1972; Fielitz and Bardack 1992; Grande and Bardack 1996; Friedman 2001a; Stewart and Friedman 2001; Fielitz and Cornett 2002; Friedman 2004; Alvarado-Ortega et al. 2006), but sharks (Friedman 2001a; Hamm 2003), plesiosaurs (Welles and Slaughter 1963; Friedman 2001a, 2004), mosasaurs (Bell and Polcyn 1995; Polcyn and Bell 1996; Jacobs et al. 2005a, 2005b), dolichosaurs (Jacobs et al. 2005a, 2005b), elasmosaurs (Welles 1949; Shuler 1950), lizards (Bell et al. 1982; Friedman 2004), and turtles (Friedman 2001a, 2004) are additionally recorded. More unique finds comprise pterosaurs and crocodiles among other marine fauna noted by Schneider and Ruez (2001). Bone-bearing coprolites useful in paleoecological analyses are documented by Friedman (2001a, 2004).

Austin Chalk
(Upper Cretaceous: Coniacian - Santonian - Campanian)

Lithology: Hard lime mudstone to soft chalk, occasionally cross-stratified. Minor forams and *Inoceramus* prisms in mostly microgranular calcite. Sparsely glauconitic; limonite and pyrite common. Locally highly fossiliferous. Deposition occurred during maximum flooding on the craton (Hovorka 1996b), allowing for higher preservation of offshore species compared to lower stratigraphic layers. Thickens to the southwest. Thickness about 177 m (580 ft) (Barnes 1977).

Fossils: Inoceramids and ammonites are reported near the Pecos River by Freeman (1961). Sharks (Bilelo 1969; Bowman 1971), a variety of fish (Price 1931; Springer 1957; Bardack 1965; Bardack 1968; Willimon 1973; Fielitz and Cornett 2002), protostegid turtles (McNulty and Slaughter 1964), ophiuroids (Clark 1959), free-floating crinoids (Marks 1952), brachiopods (Marks 1952; Young and Marks 1952), ammonites (Young and Marks 1952; Clark 1960; Young 1963), and diverse bivalves (Stephenson 1929; Marks 1952; Young and Marks 1952) are documented elsewhere in Texas. Microfauna include ostracodes (Crane 1965), isopods (Bowman 1971), and forams (Grice 1948; Young and Marks 1952; Gimbrede 1962; McNulty 1964; Clark and Bird 1966; Kariminia 2004) occasionally preserved in algal reefs (Johnson 1944). Nannofossils among other fauna listed here are additionally noted by Gale et al. (2008) in northern Texas. Evidence of gastropod predation on ostracodes is described by Maddocks (1988).

Table 1. Chronological listing of geologic formations and paleontological resources noted within Amistad National Recreation Area.

Geologic Period	Formation	Fossils Reported Within AMIS
Quaternary Holocene Epoch	**Alluvium:** gravel, sand, silt, clay, organics	none
Quaternary Holocene and/or Pleistocene Epoch	**Cultural Deposits:** middens, hearths, burials, and more	mammals, birds, reptiles, amphibians, fish, mollusks, plants, pollen, middens, and more based on archeological records
Quaternary Pleistocene Epoch	**Fluviatile Terrace Deposits:** gravel, sand, silt, and clay	none
Quaternary and/or Neogene Pleistocene and/or Pliocene Epoch	**Uvalde Gravel:** caliche-cemented gravel	none
Late Cretaceous	**Boquillas Formation:** mostly shale, some siltstone and limestone in lower parts	unknown
	Buda Limestone: fine-grained limestone, bioclastic or argillaceous	bivalves, gastropods, cephalopods, echinoids, burrows
	Del Rio Clay: calcareous and gypsiferous, iron oxide minerals common, *Ilymatogyra arietina* masses	bivalves, gastropods, forams, borings, burrows
Early Cretaceous	**Salmon Peak Limestone:** granular rudist limestone (upper part), *Globigerina* mudstone (lower part)	bivalves, gastropods, echinoids, burrows
	Devils River Limestone: limestone, dolomite, biosparite, lime mudstone	bivalves, gastropods, burrows

Cultural Deposits

Units mapped at AMIS for the Cenozoic include the Uvalde Gravel of the Pliocene or Pleistocene, fluviatile terraces of the Pleistocene, and organic-bearing alluvium from the Holocene. These deposits are not usually regarded as fossiliferous; however, cultural resources in the AMIS vicinity are often uncovered in terrace and alluvial sediments (Anderson 1974). This publication considers cultural deposits as a separate entity. Most archeological reports do not refer to specific geological units as the source of cultural material. Such deposits may include middens, hearths, and burials, that are frequently preserved in rock shelters carved out of fossiliferous limestone from the Cretaceous.

Excavations prior to the construction of Amistad Dam unearthed a plethora of Quaternary fauna and flora presently cataloged in AMIS archeological collections. These specimens could be considered in part paleontological resources despite their cultural context; however, fossils from Cueva Quebrada and Conejo Shelter are the only Quaternary remains in paleontological collections at AMIS (Santucci et al. 2001). Plentiful bones, shells, plant fibers, and pollen samples of archeological collections are nonetheless important in paleoecological analyses as mentioned by Santucci et al. (2007). Fossiliferous fire-cracked rock, pollen from numerous rock shelters, and bison specimens from Bonfire Shelter are all referred to as significant paleontological resources in the introductory chapter of this report. More Quaternary resources are described in this section as categorized by localities that may or may not fall inside park boundaries. Nevertheless, specimens recovered from all sites herein discussed are property of the National Park Service. Lack of material or uncertain provenience limit opportunities for paleoecological investigations at some localities not considered here such as Mosquito Cave (41VV215) and Castle Canyon (41VV7) (Raun and Eck 1967). Moreover, small mammals that persisted throughout the Quaternary are likely underrepresented in below taxonomic descriptions due to large sieves utilized in excavations that did not capture their remains (Raun and Eck 1967).

The following paragraphs do not represent the extent of archeological localities in the Lower Pecos region. Nor is this list comprehensive of Quaternary resources that could be considered paleontologically significant in the AMIS vicinity. However, the 14 localities described in the below section exhibit some of the best examples of the array of fauna and flora documented in the literature for the Pleistocene and Holocene. Many more publications are available that refer to other localities or additional biota recovered during early archeological investigations in and around AMIS. Further information on the hundreds of archeological sites in the vicinity can be found in an assessment by Anderson (1974). While his report does not focus on specific resources unearthed during excavations, a list of localities including formal classifications (e.g., open midden, rock shelter) and associated discoveries (e.g., pictographs, burials, perishable remains) is provided. Brief summaries of papers that might be of interest to archeologists and paleontologists are

additionally supplied in the report by Anderson. One recent inventory of AMIS cultural resources and localities by Dering (2002) is also particularly helpful as it incorporates both a cultural chronology and summary of local paleoenvironmental changes for the Quaternary. More on archeological investigations of the AMIS region can be found online at the Texas Beyond History website provided by UT Austin (http://www.texasbeyondhistory.net).

Quaternary Resources

Arenosa Shelter (41VV99)

Jurgens (2005) investigated prehistoric use of faunal resources using NPS material collected at this locality in the 1960s as it is presently inundated by Amistad Reservoir. Smaller vertebrates are represented by a variety of bony fish, amphibians, reptiles, and birds. Mammals include members of Antilocapridae, Bovidae, Canidae, Castoridae, Cricetidae, Equidae, Felidae, Geomyidae, Leporidae, Mustelidae, Mephitidae, Procyonidae, and Sciuridae. Bones and artifacts recovered from Arenosa Shelter span nearly 8,000 years of the Holocene Epoch.

Baker Cave (41VV213)

This rock opening contains a well-preserved 9,000 year old hearth filled by the remains of mammals, reptiles, fish, and plants. Douglas (1970) discusses the faunal remains found during archeological investigations at the cave. Research conducted by Sobolik (1988, 1989) on human coprolites at Baker Cave explored the dietary habits of Lower Pecos people. The most common plants observed include several members of the mustard family, sunflowers, sagebrush, sotol, and grass.

Bonfire Shelter (41VV218)

Bone deposits from the late Pleistocene and Holocene are discussed by Dibble and Lorrain (1968). Pollen from the oldest layer reflects a moist and cool climate of approximately 10,000 years ago (Bryant 1968). More on pollen records from Bonfire Shelter is provided by Hevly in Story and Bryant (1966). Cultural layers particularly rich in plant material are noted by Irving in Story and Bryant (1966) and invertebrates listed by Cheatum in the same publication consist of freshwater mollusks. Herbivores including *Bison*, *Camelops*, *Equus*, and *Elephas* are reported by Dibble and Lorrain (1968). Younger deposits include a variety of small rodents, rabbits, and other fragmentary mammalian remains. Fish and reptiles are additionally recorded. Hundreds of bones comprise potential species *Bison antiquus*, *Bison bison*, and *Bison occidentalis*. Many of these specimens are important for analyzing the "jump" method of hunting that dates back 2,600-2,700 years in this region (Dibble and Lorrain 1968). Some of the earliest evidence of human occupation in this part of Texas is represented by this locality (Dering 2002). More recent excavations of late Pleistocene deposits at Bonfire Shelter are discussed by Bement (1986).

Centipede Cave (41VV191)

Small mammal faunas and pollen assemblages recovered from this locality mirror that of the comparable yet smaller Damp Cave. Transition from moist to dry climates in what is now AMIS is documented for the Holocene based on material from these shelters by Johnson (1959, 1963).

Conejo Shelter (41VV162)

Alexander (1974) analyzed this large rock opening near Cueva Quebrada for archeological reasons including investigation of animal bones and perishable plant remains. Fauna noted in AMIS paleontological collections represent an unidentified bird, an array of fish, two turtle species, and three types of mammals. Although this locality is not found on NPS land, specimens are property of AMIS as a result of 1960s salvage operations. Conejo Shelter is one of the first localities in North America in which human coprolites were recognized, recovered, and subsequently analyzed according to the Texas Beyond History website managed by UT Austin.

Coontail Spin (41VV82)

Fauna described at Coontail Spin comprise *Canis*, *Castor*, *Erethizon*, *Lepus*, *Neotoma*, *Odocoileus*, *Ondatra*, *Sigmodon*, *Sylvilagus*, *Taxidea*, *Trionyx*, and *Urocyon*. Bones of deer, birds, and a variety of fish are further documented in these deposits that range nearly the entire length of the Holocene (Raun and Eck 1967). Extensive plant remains are reported for this shelter by Irving in Story and Bryant (1966).

Cueva Quebrada (41VV162A)

Lundelius (1984) analyzed mammalian fauna of the late Pleistocene from this small cave west of Comstock. NPS specimens are represented by mouse, rat, hare, rabbit, gopher, squirrel, skunk, ringtail, dog, fox, bear, deer, antelope, camel, horse, and bison. Although not preserved on park land, material from this locality is cataloged in AMIS paleontological collections. Most of the fragments are charred; however, it is not clear if people or spontaneous combustion of packrat middens are the cause of burned bones. The collection of material in the cave is equally mysterious as human involvement or carnivorous animals could be responsible for the mass accumulation of mammalian bones. Confirmation of anthropogenic activity in the cave could provide some of the earliest evidence of human habitation in the region (Dering 2002). Descriptions of stratigraphic layers and associated radiocarbon dates for Cueva Quebrada can be found in Lundelius (1984). NPS reports by Santucci et al. (2001) and Santucci et al. (2007) additionally review this information.

Damp Cave (41VV189)

Material recovered at this cave consists of pollen, shells, burned rock fragments, plant parts, animal bones, and aboriginal artifacts. Palynological specimens indicate the presence of *Agave*, *Celtis*, *Ephedra*, *Quercus*, *Prosopis*, and *Pinus* among other flora in the region represented by families Cactaceae, Chenopodiaceae, Compositae, Cupresaceae, Gramineae, Liliaceae, and Malvaceae (Johnson 1959, 1963).

Devil's Mouth (41VV188)

Bryant and Larson (1968) utilized fossil pollen from this presently inundated terrace locality to explore changes in late Quaternary vegetation. Taxa are represented by *Abronia*, *Acacia*, *Agave*, *Alnus*, *Celtis*, *Chenopodium*, *Ephedra*, *Gaura*, *Jussiaea*, *Liquidambar*, *Maclura*, *Mammillaria*, *Procopis*, *Pinus*, *Opuntia*, *Typha*, as well as Euphoribiaceae, Liguliflorae, and Liliaceae. More information regarding paleobotanical records and pollen analyses can be found in chapters by both Bryant and Irving in Story and Bryant (1966). Vertebrate remains are not abundant at Devil's Mouth; however, concentrations of deer, rabbit, and gophers *Cratogeomys*, *Geomys*, and *Thomomys* are noted (Raun and Eck 1967). Diverse freshwater mollusks are reported in numerous layers by Cheatum in Story and Bryant (1966).

Devils Rockshelter (41VV264)

This shallow shelter lies slightly south of the preceding locality underneath Amistad Reservoir. It has a similar composition of alluvial flood deposits yielding numerous artifacts; fauna and flora are less commonly preserved except for a variety of mollusks analyzed by Cheatum in Story and Bryant (1966). Because fragmentary pollen is difficult to identify, *Opuntia* and *Pinus* are the only documented genera from Devils Rockshelter as per Bryant (Story and Bryant 1966). Although faunal remains are present, paleoecological interpretations are limited (Raun and Eck 1967).

Eagle Cave (41VV167)

Fauna recovered from this locality are similar yet less diverse than the assemblages at Coontail Spin (Raun and Eck 1967); however, *Bison*, *Citellus*, *Geomys*, and *Mephitis* comprise additional specimens recorded for this cave. Most of the bones from Eagle Cave represent younger animals (Raun and Eck 1967). Holocene pollen records are discussed by McAndrews and Larson in Story and Bryant (1966) and diverse plant remains are described for Eagle Cave by Irving in Story and Bryant (1966). Mollusks consist of freshwater mussels (*Amblema*) and land snails (*Bulimulus*) as listed by Cheatum in Story and Bryant (1966).

Hinds Cave (41VV456)

Williams-Dean (1978) analyzed various components of human coprolites from Hinds Cave. Palynological investigations yielded zoophilous pollen (dispersed by insects, birds, and bats) and anemophilous pollen (dispersed by the wind). Fruit, seeds, and bones comprise most macrofossils. Flora encompass *Agave*, *Allium*, *Amaranthus*, *Celtis*, *Cenchrus*, *Chenopodium*, *Dasyliron*, *Diospyros*, *Juglans*, *Opuntia*, *Prosopis*, *Sporobolus*, *Vitis*, *Yucca* and members of the Tribe Paniceae. Dering (1979) further described pollen and plant macrofossils uncovered at Hinds Cave. Fauna classified to at least the genus level comprise *Aplodinotus*, *Citellus*, *Colinus*, *Lepus*, *Neotoma*, *Onchomys*, *Ondatra*, *Peromyscus*, *Procyon*, *Rana*, *Sigmodon*, *Sylvilagus*, *Urocyon*, and *Zenaidura* (Williams-Dean 1978). Other unidentified fragments of mammals, birds, reptiles, and fish are also noted. Supplemental analyses of older Holocene coprolites by Stock (1983) yielded similar macrofossil components and Lord (1984) reported the overall occurrence of at least 60 vertebrate taxa based on his zooarcheological investigations. Fur, scales, land snails, mussels, and arthropods including a grasshopper, millipede, beetles, lepidopteran, and dipteran larvae are additionally mentioned among these publications.

Parida Cave (41VV187)
Quaternary cultural resources are preserved in the form of domestic midden deposits at Parida Cave. Alexander (1970) reported chipped stone, bone, shell, and antler artifacts from 1967 excavations. Other faunal and floral remains, coprolites, and pollen are additionally recorded.

Zopilote Cave (41VV216)
The number of bones unearthed at this burned rock midden off of Seminole Canyon is small compared to other sites examined by Raun and Eck (1967). The concentration of rabbit material is high relative to poor representation of deer, turtles, and fish. It appears that indigenous people of the middle Holocene did not bring many aquatic food items to this rock shelter (Raun and Eck 1967). However, an array of fragmentary plants is noted for this locality by Irving in Story and Bryant (1966).

Paleontological Resources Inventory

AMIS fossils are represented by plants, invertebrates (including microfossils), vertebrates, and trace fossils (ichnofossils). These can be found in the field and in NPS collections held at park headquarters, Texas Archeological Research Laboratory (TARL), and the Vertebrate Paleontology Laboratory (VPL) at UT Austin. Quaternary specimens are often the result of archeological excavations; most resources are not curated in paleontological collections. Such specimens are considered paleontological resources, however, and are consequently included in this inventory. Because more research has commenced on Quaternary specimens compared to Cretaceous fossils, identification of Pleistocene and Holocene material is more comprehensive than that of Cretaceous. Taxa are compiled to the genus level in Appendix A based on review of the literature, park collections, and field observations. Names gleaned from published research may not represent current classification schemes. Localities reviewed in the *Cultural Deposits* chapter are the focus of Quaternary genera presented in Appendix A. This review is likely incomplete for local fauna and flora of archaeological records; nonetheless, Appendix A offers a basic understanding of generic diversity in the vicinity of AMIS for the Quaternary.

Fossil Plants

Fruit, seeds, pollen, wood, and plant fibers are abundant in cultural deposits of southwestern Texas. Although specimens are cataloged as archeological resources, paleobotanical records are often utilized in paleontological analyses (e.g., paleoclimate studies). Taxon lists are provided in Appendix A based on archeological sites discussed in the chapter on cultural resources and NPS specimens collected at Fate Bell Shelter of nearby Seminole Canyon State Park and Historic Site. Nearly 50 plant families are recorded for Bonfire Shelter, Centipede Cave, Coontail Spin, Damp Cave, Devil's Mouth, Devils Rockshelter, Eagle Cave, Hinds Cave, and Zopilote Cave by Johnson (1959, 1963), Bryant, Irving, Hevly, and McAndrews and Larson in Story and Bryant (1966), Bryant and Larson (1967), Williams-Dean (1978), and Stock (1983). Baker Cave, Conejo Shelter, and Parida Cave similarly contain Quaternary pollen and plant remains, but are not listed in Appendix A as they were not identified in the literature available at the park.

Fossil plants are not recorded in Cretaceous deposits except perhaps for algae mentioned by Jones (1993). He notes the presence of algae in all AMIS strata; however, no further information is provided. It is not clear if specimens are actually observed inside park boundaries or if algal occurrence is solely supported in these units based on literature comparisons. Moreover, Jones (1993) does not comment on the variety of algae reported. Because algae are not only represented by plants, paleobotanical reports for the Cretaceous cannot yet be confirmed at AMIS.

Fossil Invertebrates

Fragments of Upper Paleozoic rock found inside AMIS boundaries at Diablo East contain abundant crinoids, gastropods, bivalves, and bryozoans. This material did not originate near the park; nonetheless, such fossils may be useful in interpretive programs. Taxa are not listed in Appendix A as Paleozoic rock is not found in the AMIS vicinity and specimens are not curated in paleontological collections. However, one fossiliferous slab of Upper Paleozoic material is informally stored at the park in an education collection.

Mesozoic strata are dominated by marine invertebrates. Microfossils consist of ostracodes and planktonic and benthic forams for all Cretaceous deposits in the park (Smith and Brown 1983; Jones 1993). Mollusks are represented by an array of bivalves, cephalopods, and gastropods in Lower and Upper Cretaceous units as documented by Smith and Brown (1983), Jones (1993), and Visaggi (2006). Large rudistid bivalves are particularly abundant; however, identification is lacking for park specimens. Fieldwork on rudists in and around AMIS by Kerans (2002) and Scott (2002a) may assist in future identification. Publications by Filkhorn et al. (2005) and Scott (2007) may be additionally helpful. Fossil echinoids, ophiuroids, corals, and bryozoans are similarly found inside the park as noted by Jones (1993) and Visaggi (2006). Taxa compiled in Appendix A are based on literature review, collection updates, and field observations.

Quaternary deposits in the vicinity additionally contain invertebrate specimens. Fragmentary arthropods, land snails, and mussels are noted for Hinds Cave by both Williams-Dean (1978) and Stock (1983). Diverse non-marine mollusks are reported by Cheatum in Story and Bryant (1966) for four localities: Bonfire Shelter, Devil's Mouth, Eagle Cave, and Devils Rockshelter. Quaternary shells recovered in archeological investigations are held at TARL despite their potential utility in paleontological research. Shells are recorded at other localities including Damp Cave and Parida Cave; however, only genera listed by Cheatum in Story and Bryant (1966) are available in Appendix A.

Fossil Vertebrates

Mesozoic strata at AMIS may contain pieces of marine vertebrates; however, none are presently known inside park boundaries.

Quaternary deposits are the only documented source of vertebrate material within the park. Bones, skin, and other remains from the Pleistocene and Holocene are chiefly held in NPS collections at TARL in Austin. Birds, fish, amphibians, reptiles, and an assortment of mammals are frequently found at archeological sites including the fourteen localities mentioned in the chapter on cultural deposits in this report. Fossils from Cueva Quebrada and Conejo Shelter are the only Quaternary specimens cataloged as paleontological resources at AMIS. Taxa listed in Appendix A for many of the above rock shelters are compiled based on chapters by Raun as well as Lorrain in Story and Bryant (1966), Raun and Eck (1967), Dibble and Lorrain (1968), Williams-Dean (1978), Lundelius (1984), and Jurgens (2005).

Trace Fossils

Observations of trace fossils (also called ichnofossils) are common in all Cretaceous units at AMIS according to Smith and Brown (1983), Jones (1993), and Visaggi (2006). Trace fossils such as *Skolithos*, *Planolites*, and *Thalassinoides* reported by Zahm et al. (1995a,1995b) in nearby IBWC cores and similar ichnofossils viewed inside park boundaries suggest the presence of active crustaceans and worms in the Mesozoic. Burrows are particularly abundant in the iron-stained Del Rio Clay (Jones 1993; Visaggi 2006) and evidence of boring sponges (e.g., *Cliona*) is frequently noted on the oysters therein (Visaggi 2006). Ichnogenera (trace fossil genera) viewed in Cretaceous strata at AMIS are listed in Appendix A.

Trace fossils of the Quaternary no doubt include packrat middens found in rock shelters; however, because trace fossil genera are not listed in the literature reviewed for this report, they do not appear in Appendix A.

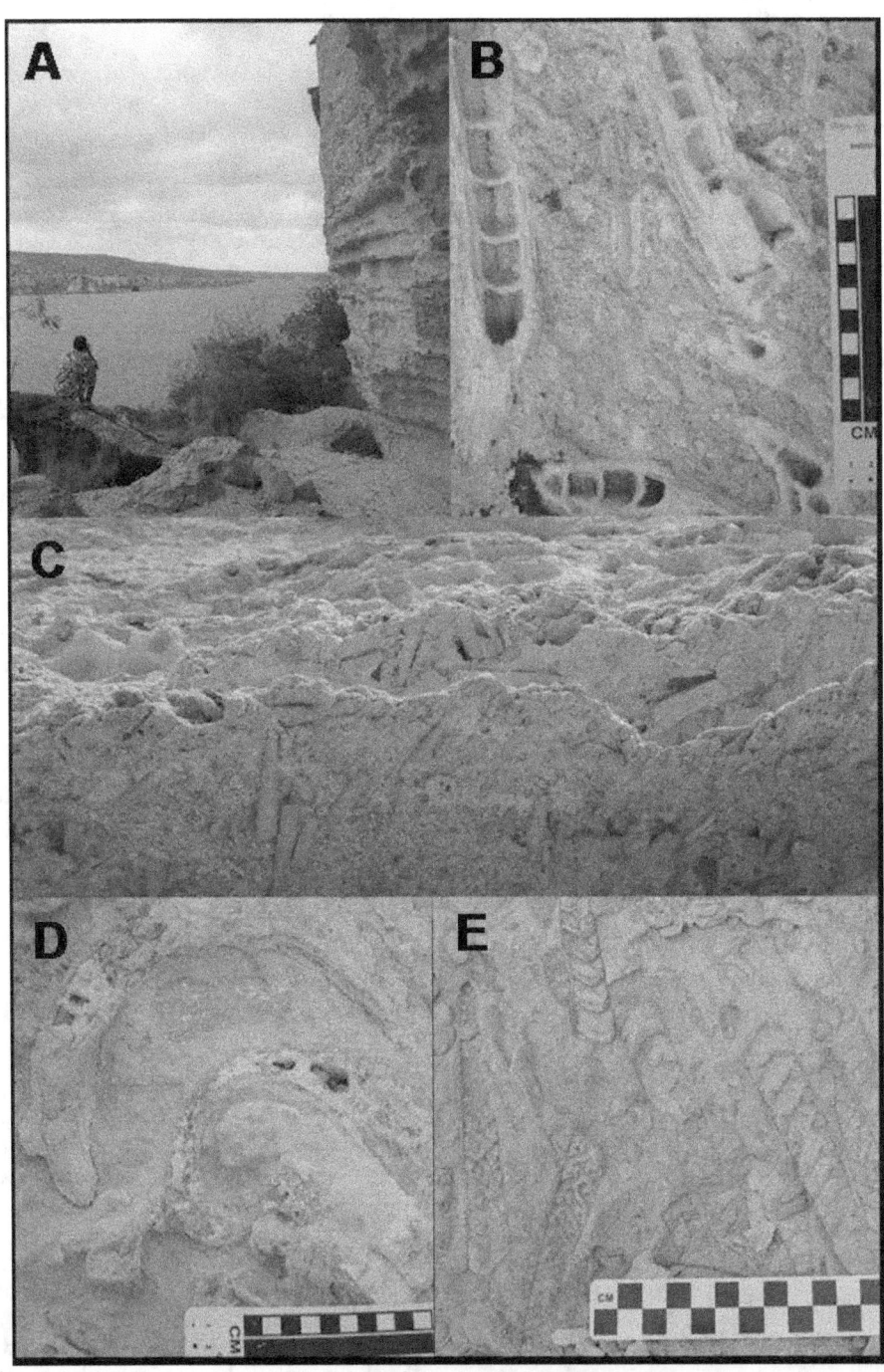

Figure 4. Salmon Peak Limestone. A) Fossiliferous bluffs on the Rio Grande. Jack Johnson for scale. B) Exposed interiors of rudistid bivalves. C) Terraces carved by aeolian erosion. D) Rudistid bivalves. E) Vertical cross-sections of *Nerinea*. Photos by Angel Johnson (A), Christy Visaggi (B, D), and Jack Johnson (C, E).

Localities

Field investigations completed during March 2006 covered 24 in-park localities, eight Val Verde County road cuts, and two neighboring sites managed by the Shumla School. The opportunity to visit a newly recognized fossiliferous locality at AMIS as well as the adjacent Seminole Canyon State Park and Historic Site occurred in January 2007. Documentation of paleontological localities has not been conducted previously for AMIS. This is the first attempt to identify fossiliferous areas on National Park Service land.

Field efforts concentrated on Cretaceous deposits; accessibility (e.g., inundated localities, caves found on private land) restricted inspection of cultural localities bearing Quaternary paleontological resources. These specimens are heavily documented, however, as a result of archeological salvage operations prior to the formation of the reservoir. Lack of knowledge regarding Cretaceous fossils in the park guided the decision to primarily focus on Cretaceous strata in the field. Exploration of all geological formations mapped inside park boundaries commenced either in the park or along nearby road exposures if not accessible at AMIS. Geological maps, existing rock shelters, staff recommendations, popular park areas, interpretation possibilities, monitoring concerns, ad hoc observations, and a desire for good spatial coverage contributed to the choice of localities analyzed.

Appendix B contains a map of several localities reviewed in the AMIS vicinity as part of this investigation. Discussion of park localities is summarized based on four geographical regions; information is limited in this version of the report as the park is working to actively manage its paleontological resources. Other localities are discussed in more detail as they are not found on NPS land. Only public road cuts are plotted on geologic maps in Appendix C as modified from Barnes (1977). Maps by Barnes were used to identify formations in the field and may not reflect new interpretations of the region based on sequence stratigraphic analyses. Fieldwork focused on recognizing geological units, identifying characteristic fauna, collecting unique specimens, and assessing the condition of paleontological resources. Options for future interpretive programs and management plans are additionally considered, but are not presented here in an effort to protect paleontological localities inside park boundaries.

Of the 25 localities analyzed at AMIS, six are considered especially significant and recommended for reporting through the National Park Service GPRA Goal 1a9A. These localities require additional management strategies as explained in the *Management and Protection* chapter of this report.

Park Localities

I. Diablo East, Evans and California Creek, and San Pedro Area

This vicinity is largely composed of Lower Cretaceous Salmon Peak Limestone; however, Upper Cretaceous Del Rio Clay and Buda Limestone are also mapped in this region. Land near the Amistad Dam along the Rio Grande may additionally contain surface layers of Cenozoic material.

Fossils commonly found in this part of the park include regular and irregular echinoids, a variety of crystalline cross-sections of mollusks, high-spired snails, large rudistid bivalves, *Neithea* scallops, *Ilymatogyra arietina* oysters, *Cribratina texana* forams, and an assortment of unidentified fragments of either molluscan, anthozoan, or algal affinities. Oyster specimens are often leached or partially replaced by iron sulfides resulting in reddish-purple colored shells. These shells are frequently bored by encrusting sponges (e.g., *Cliona*). Worm burrows are additionally noted in Lower Cretaceous units. Most of the bedrock is covered by vegetation, soil, or lichen film; abundant empty shells of modern *Corbicula* on the ground indicate higher water levels are not uncommon.

Several paleontological resources are also found in the form of fire-cracked rock (FCR) near archeological localities. These fossils derived from areas known as burned rock middens (BRM), which were formerly utilized as cooking centers by native peoples of the Lower Pecos. Fossils characteristic of these thermally altered rocks suggest a stratigraphic assignment to the Del Rio Clay. The preference for fossiliferous rocks in hearth features is discussed in the *Introduction* chapter of this report.

II. Rio Grande and Pecos Confluence

This part of the park primarily contains exposures of the Lower Cretaceous Salmon Peak Limestone; however facies change to the in part laterally equivalent Devils River Limestone occurs in this vicinity. Common fossils include rudistid bivalves and *Nerinea* (high-spired gastropods commonly found in rudistid reefs of the Mesozoic). Scallops, clams, and burrows are occasionally noted. Some fossils are severely affected by post-burial chemical alteration. Erosion of fossiliferous bluffs often exposes the interior of molluscan shells in this region, particularly useful for identification of *Nerinea* as they exhibit a characteristic pattern in cross-section. The interior of rudistid bivalves commonly reveals multiple chambers filled by large calcite crystals. Fossils affected by post-burial processes are delicate and not conducive to collection; however, such differential erosion of limestone offers beautiful views of fossiliferous ledges (Fig. 4). Archeological localities in this part of the park additionally contain paleontological resources. Domestic midden deposits include material such as plant fragments and fish bones utilized by native peoples of the Lower Pecos.

III. Hwy 90 at Spur 406 and West to the Lake

Numerous geological formations are mapped in this portion of Val Verde County. Much of the shoreline is composed of Salmon Peak Limestone; however, exposures of the Del Rio Clay and Buda Limestone are additionally present in this part of the park. Fossils found in this vicinity include *Ilymatogyra arietina*, some of which exhibit a reddish-purple color due to iron alteration. Several oyster valves additionally contain holes characteristic of the boring sponge *Cliona* (Fig. 5). Other local fossils include crystalline cross-sections from a variety of mollusks, *Neithea*, agglutinated forams, and burrows made by worms and crustaceans. Lichen film and vegetation restricts exposed sections in this region.

IV. Shumla Bend and West Along Rio Grande

Rock exposures in the western edge of the park are limited and difficult to access along the Rio Grande. These cliffs are mostly composed of Lower Cretaceous Devils River Limestone. Fossils include molds and cross-sections of mollusks including high-spired snails, *Neithea*, and oysters. Canyon exposures are not likely to yield abundant fossils as rocks are very hard and greatly affected by weathering and lichen growth.

Other Localities

Shumla School

The Shumla School, a non-profit education center that focuses on human use of materials, land, and art, manages 648 ha (1,600 acres) in Val Verde County. Their programs primarily explore ancient cultures of the Pecos River region; however, paleontological interpretation is possible as nearly the entire area is underlain by fossiliferous bedrock. Members of the Shumla School are eager to incorporate paleontology into their curriculum and in fact a program for middle school children commenced a few months after investigation of the below localities. Interpretive programs at AMIS could equally benefit from land and resources available through the Shumla School as staff at both locations are open to partnership options.

Shumla School property northwest of Comstock contains a wealth of Lower Cretaceous fossils. Although the region is supposedly underlain by Devils River Limestone, paleontological and lithological evidence reveal striking similarities to shelter exposures of Salmon Peak Limestone within AMIS. This locality is subject to high rates of erosion and crystalline fossils consisting mostly of *Nerinea* and rudistid bivalves are extremely fragile. The shelter is deep in a valley west of the Pecos River; access requires a challenging hike on an unmarked path. Fossil exposures are beautiful; however, collection and interpretation opportunities through the Shumla School might be limited by remote location and delicate preservation.

The second paleontological locality utilized by the Shumla School near Comstock is privately owned by local benefactors of the organziation. The property is mostly mapped as Salmon Peak Limestone in canyons and undivided Del Rio Clay and Buda Limestone at higher elevations. Canyon exposures of Salmon Peak Limestone are difficult to access and are not very fossiliferous (except for random small patches of broken fire-cracked rock containing oysters of the Del Rio Clay). Sections of the Del Rio Clay and Buda Limestone, however, contain abundant paleontological resources nearby. Ground exposures and rock fragments are full of *Neithea*, *Ilymatogyra*, and *Cribratina* (Fig. 5) characteristic of the Del Rio Clay. Other fossils include a variety of bivalves, gastropods, cephalopods, irregular

echinoids, and burrows. Specimens are commonly found on the ground having eroded out of the rock; stratigraphic origin either belongs to the Del Rio Clay or Buda Limestone. Small pieces of thermally altered rock are additionally reported. This locality is a wonderful resource for the Shumla School. Although the area requires a prolonged drive on slightly rugged ground, it is accessible, safe, and contains plentiful fossils that can be identified easily as assisted by a basic fossil guide due to good preservation. The locality is suitable for all ages. Several specimens uncovered as part of this investigation are now in the collections at the Shumla School for future use in interpretive programs (Fig. 5).

Seminole Canyon State Park and Historic Site

Amistad National Recreation Area and Seminole Canyon State Park and Historic Site share a boundary southeast of the Pecos River. Marine fossils of the Lower and Upper Cretaceous are mapped in this vicinity and interpretive panels found outside the visitor center convey both paleontological and geological information for the region. Canyon erosion, rock shelter formation, and microbial interactions are discussed under the heading "Sculpted by Nature." Limestone cliffs in southwestern Texas are often stained dark gray or black. Originally believed to be of inorganic origin (J. Labadie, pers. comm.), researchers now attribute such stains to lichens and other microbes (Kaluarachchi et al. 1995). These microorganisms additionally contribute to the existing array of physical forces of erosion that wear on canyon walls.

The second panel, "A Journey Through Time," provides information on geologic history, stratigraphy, and fossils frequently found in these deposits. Depictions of warm seas from the Cretaceous exhibit an array of marine life that once inhabited southwestern Texas nearly 100 million years ago. Diverse mollusks are displayed either in painting or as plaster casts attached to the sign. Processes of fossilization are briefly reviewed. Photos of local road cuts are utilized alongside formation descriptions in the stratigraphic section that incorporates the Devils River Limestone, Del Rio Clay, Buda Limestone, and Boquillas Formation. Most of this park is actually underlain by the Salmon Peak Limestone as per Barnes (1977), but a facies shift to the Devils River Limestone occurs nearby slightly northwest of park boundaries. Other small errors in the sign regard the discussion of swimming dinosaurs (mosasaurs are classified as ancient reptiles and are not dinosaurs) and the use of obsolete names for some fossils (e.g., oysters of the Del Rio Clay belong to the *Ilymatogyra* lineage and are no longer considered of the genus *Exogyra*).

Fossils from the Salmon Peak Limestone and Del Rio Clay are documented at Seminole Canyon State Park and Historic Site. Specimens include *Ilymatogyra arietina*, high-spired gastropods, cross-sections of rudistid bivalves, and smaller clams. The opportunity for joint paleontological programs between staff at Amistad National Recreation Area and Seminole Canyon State Park and Historic Site could be explored in the future.

Highway Exposures

Eight road cuts in the AMIS vicinity are described as part of this investigation. These areas are not inside park boundaries, but may be useful for park staff seeking accessible locations that clearly show differences in local stratigraphic units.

The Lower Cretaceous Salmon Peak Limestone is not visited at highway road cuts as it is abundantly exposed inside the park. However, Devils River Limestone is explored along the access road to the boat launch at the Pecos River (Locality RC7). Although this section of road past the top of the hill is beyond NPS boundaries, crystalline fossils are displayed in rocks on the north side of the road. Most fossils are viewed in cross-section; gastropods, clams, and rudistid bivalves are the most common paleontological resources observed. East of the Pecos River bridge along Rt. 90 is another exposure of Devils River Limestone that additionally contains layers of the overlying Del Rio Clay and Buda Limestone (Locality RC5). This outcrop provides a clear distinction between Lower and Upper Cretaceous stratigraphic units. Fossils are noted most in rocks on the ground.

Upper Cretaceous units are further documented at a number of localities near AMIS revealing portions of the Del Rio Clay, Buda Limestone, and Boquillas Formation. Several miles north of Spur 406 off of Rt. 90, fossils are observed in the Del Rio Clay (Locality RC4). The contact between the Del Rio Clay and overlying Buda Limestone (Fig. 6) is clearly exposed here and at another site a few miles north along the highway. The clay is mustard brown and contains *Ilymatogyra*, while the Buda is of lighter color and either crumbly or blocky in appearance. Heading south along Rt. 90 near the farthest crossing of Evans Creek, exposures of the overlying Boquillas Formation (Fig. 6) are noted on both sides of the road (Locality RC3). Beautiful orange and red wavy patterns are seen in brittle outcrops of this Upper Cretaceous unit. The formation contains fossils elsewhere, but this locality appears relatively barren. Tepee structures are observed, reflecting movement of rock pressured by surface caliche formation.

The Eagle Ford Group and Austin Chalk (Upper Cretaceous) as well as Uvalde Gravel (Neogene-Quaternary) are investigated at local road cuts as their exposure inside the park is minimal, questionable, or inaccessible. Slightly north of the Rt. 277 and Rt. 90 intersection are fossil-bearing outcrops of the flaggy Eagle Ford Group (Locality RC1). Fossils include crystalline cross-sections and fragmentary bivalves, some showing signs of iron oxidation. One allochthonous (not *in situ*) older slab of *Ilymatogyra arietina* is also noted on the side of the road. Eagle Ford deposits are additionally observed south of the reservoir bridge heading north on Rt. 277 (Locality RC8). Although the Eagle Ford is not fossiliferous at this locality (similar to its nearby Boquillas Formation counterpart), underlying crumbly Buda Limestone contains several shell cross-sections, crystalline fossils, and burrows.

The Austin Chalk, the uppermost Cretaceous unit in the vicinity, is observed beyond Langtry on Rt. 90 (Locality RC6). Macrofossils are not common on surface exposures, but several shell fragments and burrows are reported. The Austin Chalk may be renowned for its diverse and abundant fauna; however, sections in this part of Texas appear to be void of large macrofossils.

Lastly, north of the Rt. 277 bridge is the caliche-cemented Uvalde Gravel (Locality RC2). These cobble-filled deposits are not usually regarded as fossiliferous. The only local fossils recorded from recent deposits are those found in rock shelters scattered throughout the region as described earlier in this report.

Figure 5. Sample of fossils from AMIS and Shumla School localities. A) Nautiloid collected for the Shumla School. B) *Ilymatogyra arietina* oysters collected for Amistad National Recreation Area. C) *Cribratina texana* (agglutinated uniserial forams) noted near the Shumla School. D) *Cliona*-like boreholes in Del Rio Clay oysters documented inside the park. Photos by Jack Johnson (A, B, D) and Shannon Garard (C).

Figure 6. Outcrops visible along Rt. 90 near Amistad National Recreation Area. Left: Contact between Del Rio Clay (lower tan layers) and Buda Limestone (upper gray layers). Right: Flaggy (layered) Boquillas Formation rocks. Photos by Christy Visaggi.

Collections and Curation

Museum Collections

The Amistad National Recreation Area museum collection is one of the ten largest collections in the National Park Service and has a recently completed Collection Management Plan (Labadie et al. 2005). Most specimens are the result of 1960s excavations conducted by the Texas Archeological Salvage Project (TASP) before the construction of Amistad Dam. The majority of cataloged objects are cultural artifacts; however, natural history specimens including diverse paleontological resources are not uncommon. Fossils chiefly consist of Cretaceous marine invertebrates and Quaternary fauna and flora found in rock shelters during TASP salvage work. The primary repository for AMIS collections is the Texas Archeological Research Laboratory (and adjacent Vertebrate Paleontology Laboratory) at the University of Texas at Austin. A small number of paleontological resources are stored in a unit adjacent to park headquarters in Del Rio.

Most Cretaceous marine fossils accessioned in AMIS collections are housed in Del Rio. Smokey Lehnart, former AMIS park naturalist, collected the majority of these specimens at some unknown point over 25 years ago. He did not record locality or stratigraphic information for these fossils; many lacked proper taxonomic identification as well. Specimens are cataloged in ANCS+ despite this poor documentation. This research effort re-examined fossils from Smokey's collection providing updates on age, formation, condition, and identification wherever possible. Although more than half of the collection contains incomplete specimens, most are in good or fair condition. The senior author also added several new fossils to the collections as a result of March 2006 fieldwork. Taxa presently cataloged in the paleontological collections at AMIS headquarters are listed in Table 2. Two photos of museum specimens housed in Del Rio are additionally provided (Fig. 7).

For the current study, investigation of Quaternary paleontological resources relied on collection records, literature review, and personal communication, whereas field research focused on Cretaceous units. Temporal constraints limited access to collections held in Austin; AMIS staff aided in obtaining data on specimens cataloged as paleontological resources, including hundreds of Pleistocene bones from Cueva Quebrada and Conejo Shelter. These rock shelters are not inside park boundaries, yet AMIS maintains ownership of specimens from these localities as a result of early salvage excavations. Taxonomic lists for both Cueva Quebrada and Conejo Shelter are provided in Table 3. These fossils are described in greater detail in a publication by

Lundelius (1984) and species names suggested in that report (but not listed in AMIS records) can be found in NPS publications by Santucci et al. (2001, 2007).

NPS collections recovered from Arenosa Shelter, Bonfire Shelter, Baker Cave, Centipede Cave, Damp Cave, Eagle Cave, Parida Cave, Hinds Cave, Zopilote Cave, Coontail Spin, Devil's Mouth, and Devils Rockshelter among other localities

similarly exhibit a range of vertebrates, invertebrates, plants, microfossils, and trace fossils and from the Quaternary. These resources are important indicators of human diet, hunting behavior, cooking methods, and changes in paleoclimate and vegetation, but specimens are usually curated in the archeological collections instead. More information on Quaternary fossils and localities in the AMIS vicinity can be found in the cultural resources section of this report and in the NPS publication on paleontological resources of the Chihuahuan Desert Network (Santucci et al. 2007).

Museum Repositories

The primary repository for archeological collections in the state of Texas is the Texas Archeological Research Laboratory (TARL), located on the J.J. Pickle Research Campus of the University of Texas at Austin. Although the Vertebrate Paleontology Laboratory (VPL) adjacent to TARL manages the bulk of paleontological resources for the park, Quaternary fauna and flora classified as archeological specimens can be found in both Room 19 and Building 33 of TARL. Several paleontological specimens are additionally stored at AMIS headquarters. These facilities and their collections are briefly reviewed in the following paragraphs; more information is available in the AMIS Collection Management Plan (Labadie et al. 2005).

Texas Archeological Research Laboratory (TARL)

Bones, shells, wood, plant fibers, and animal skin are stored in Room 19. This humidity- and climate-controlled storage space contains the majority of cataloged fauna and flora recovered from Quaternary rock shelters as a result of TASP operations. Unprocessed faunal material of lesser value is held in Building 33. Humidity and climate are not controlled in this large and open space accessible to cars. Most samples remain exposed in paper bags or cardboard boxes; pest management is a major concern in Building 33. Specimens at TARL are protected by locks, alarms, and a restricted access policy in most cases. The relationship between TARL and AMIS is in good standing; however, further communication and agreements regarding loan policies and research requests according to NPS guidelines should be pursued (Labadie et al. 2005).

Vertebrate Paleontology Laboratory (VPL)

This is the storage location for paleontological resources uncovered at Cueva Quebrada. Other AMIS collections held at this facility include specimens from Arenosa Shelter, Centipede Cave, Coontail Spin, Damp Cave, Devil's Mouth, Eagle Cave, Mosquito Cave, and Parida Cave. Samples recovered from Conejo Shelter and Bonfire Shelter are now located at TARL. Material is either uncataloged or cataloged in MS Access; ANCS+ is not used by VPL personnel. Specimens are stored in drawers, on open shelves, or in boxes that may be found upstairs, downstairs, or in the basement as is common for large or unprocessed samples. There are no climate records for VPL, but humidity and climate are purportedly controlled in most areas except for the upstairs warehouse containing cataloged and uncataloged bone material. Pest monitoring is reported ad hoc; high levels of security are supposedly in effect. One goal of the AMIS Collection Management Plan (Labadie et al. 2005) is to move park specimens from VPL to TARL for greater consistency in organizing material until a larger facility is available exclusively for AMIS collections. This could additionally alleviate many of the concerns regarding improper storage of specimens at VPL as well as inconsistencies in documentation. Entering data into ANCS+ according to NPS guidelines in the meantime is another objective mentioned by park staff as it is presently difficult to obtain information from VPL. A PMIS (Project Management Information System) statement could be submitted to acquire funds for backlog cataloging (G. Bell, pers. comm.). Taxon lists for many of the archeological localities mentioned above are available online however, and can be accessed through the VPL website hosted by UT Austin. Yet, moving AMIS samples to TARL and entering data into the ANCS+ system remain the preferred NPS management option for the future.

AMIS Headquarters

The paleontological collections at park headquarters in Del Rio are chiefly composed of Cretaceous specimens. Temporary space is all that is currently available for storage; however, future goals include the construction of a larger building for increased museum exhibits, collection space, and perhaps lab facilities. Specimens are stored in a non-insulated, climate-controlled room. The dehumidifier has previously created problems at this storage location, but ongoing efforts to change the focus of humidity control from the work to collections area should ease humidity concerns. AMIS fossils occupy a few drawers within a single cabinet; no organization is followed except that large specimens remain at the bottom. Some fossils are extremely crystalline and require careful handling and storage. The collections at park headquarters are stored in a locked building, behind a locked door, in a series of locked specimen cabinets. Building access is only possible through the designated manager of all collections and the chief of education and resource management; keys for specimen cabinets are reserved solely for the acting collections manager (R. Slade, pers. comm.). Visitor access is not permitted unless an escort is present as is similar to procedures followed at TARL and VPL. All visitors must sign the entry log before heading into the collections building.

Management Considerations

Scope

There is no Scope of Collections Statement (SOCS) at present for AMIS. The Collections Policy Statement of 1976 is no longer accepted by the NPS as an approved document guiding museum collections. This should be replaced by a SOCS that incorporates archeological, biological, archival, historical, geological, as well as paleontological collections (Labadie et al. 2005). The former park archeologist had initiated this effort; however, to facilitate timely production of this document a Technical Assistance Request could be submitted to the Museum Services Division of the NPS in order to acquire help in writing a Scope of Collections Statement.

Archives and Photography

Little documentation exists for previous paleontological resources at AMIS as data did not accompany Cretaceous fossils in the Smokey Lehnart collection. Information from this research effort including digital copies of field notes, locality maps, and collection photos shall be filed in the archives at park headquarters. Jack Johnson, former SCA archeological intern for the park, photographed all existing paleontological specimens housed in Del Rio and several fossils in the Shumla School collections. Shannon Garard, former SCA museum intern, and Melissa Webster, former AMIS cultural resources web intern, likewise photographed fossils added to the collection by the senior author.

Obtaining digital images is an increasingly important aspect of collections management. Future paleontological research at AMIS could greatly benefit if the few Cretaceous specimens held at TARL in Austin are similarly photographed. These photographs might be of interest to the public, park employees, and visiting scientists. Joe Labadie, retired AMIS cultural resources manager, and Phil Dering, archeobotanist instructor at the Shumla School, are in support of a future online resource to increase paleontological awareness in the region; an archive of photographs to select from would be helpful in achieving this goal. Interpretive programs could equally benefit from such an archive in creating slide presentations.

Quaternary resources are heavily documented as a result of TASP operations; however, data are restricted to archeological archives in most cases. These archives likely contain useful information for Quaternary paleontologists such as the folders at VPL in Austin filled with field notes and images for a dozen archeological localities including Bonfire Shelter. Material presently located at a variety of institutions needs to be consolidated and archeological localities containing recent paleontological resources should be cross-referenced. Acquiring specimen images and historic photos for caves containing paleontological resources such as Cueva Quebrada and Conejo Shelter could be another objective of archive management in the future.

Specimen Relocation

The small number of Cretaceous specimens held at TARL in Austin could be removed to park headquarters. These fossils are property of the park and having a partial collection at a secure location far from the field area is not useful for paleontological research or interpretation. Transfer of specimens may not be feasible immediately as it may depend on availability of storage space and current responsibilities of collections staff at AMIS. However, plans for a new visitor center including additional space for museum collections could facilitate the relocation of TARL Cretaceous specimens to park headquarters in Del Rio.

Publication Specimens

Little paleontological research has occurred since the inception of the park, yet numerous publications mentioned at the beginning of this report including Coogan (1973), Smith and Brown (1983), Humphreys (1984a, 1984b), Zahm et al. (1995a, 1995b), Kerans et al. (1995), Kerans (2002), and Scott (2002a, 2002b) refer to localities in and around AMIS.

Further research is needed to confirm precise locations relative to NPS boundaries. Relevant information should be filed in park archives and documentation or relocation of fossils to current park repositories should be completed. This may include material collected by Kerans and utilized by Scott presently curated at the Texas Memorial Museum (Scott 2002b). Their samples may have been collected from sections within the park, but maps and locality descriptions in their reports are not clear enough to make this determination.

It is important that the park addresses issues of unclear ownership before accessioning or cataloging any specimens. If paleontological localities cannot be verified as being on NPS property or if materials were collected prior to establishment of the park, AMIS staff will need to negotiate concessions of ownership from those repositories currently in possession of the materials. Once questions of ownership are resolved the park can then produce accessions that specify or estimate the number of items requiring backlog cataloging. That information could be used to develop PMIS statements for future backlog cataloging funding (G. Bell, pers. comm.).

Interpretation Collection
Fossils in the collections at park headquarters are supposedly used for intermittent interpretive programs, but this could not be confirmed. There is no formal loan system or logbook that exists for such purposes at AMIS HQ (R. Slade, pers. comm.). No active natural resources manager or specialist has been employed by the park until recently (Labadie et al. 2005), likely contributing to the lack of interest in utilizing paleontological resources. Angel Johnson, Shumla School staff member and former SCA museum intern at AMIS, explored options to start a separate paleontological collection for educational purposes. This collection could provide an easily accessible set of local fossils to be stored at the visitor center for interpretation.

Limited space at park headquarters restricts the possibility for large paleontological collections; however, rare species, well-preserved specimens, and unique threatened resources should be accommodated as part of the museum collection. Museum collections should ultimately reflect the range of species and preservation found within the park. By creating an independent collection for interpretation and visitor enjoyment at the visitor center, more space could be available for collections requiring secure storage conditions. Poorly preserved specimens and those lacking informative records (e.g., several fossils collected by Smokey Lehnart) could be deaccessioned and moved to the interpretation collection as long as duplicate specimens of those species, with proper documentation, remain represented in museum collections. Specimen records should still accompany fossils in the interpretation collection for management purposes (even if not incorporated in ANCS+ alongside official museum collections). This helps avoid loss or misplacement of fossils as befell a large ammonite previously exhibited at the visitor center at Rough Canyon (J. Little, pers. comm.)

Furthermore, a loan system accompanied by a logbook documenting the occurrence of borrowed specimens is the best method in preventing loss as it holds individuals accountable. Tracking specimen whereabouts is standard practice in the management of museum collections and a check-out sheet or logbook could be a simple solution. Although paleontological resources could be utilized more frequently for educational purposes at the park, specimens could benefit from management as part of a separate collection dedicated solely to interpretation.

Further Recommendations for Researchers
Major concerns regarding management of natural history collections at AMIS include inconsistency in park and researcher communication and lack of adherence by researchers to NPS guidelines for reporting, cataloging, and curating specimens (Labadie et al. 2005). This is an issue for many parks and could ultimately lead to negative impacts for resources, loss of NPS property, loss of data, and further increases the already monumental NPS backlog cataloging problem. Although this research effort alleviated some of these concerns at AMIS by attempting to fill in the gaps for previous paleontological specimens that lacked collection records, only so much could be done in the absence of locality and stratigraphic information. Material added to park archives and collections as a result of this investigation should serve as a guide for any future paleontological research conducted at AMIS. The following recommendations are additionally offered (G. Bell, pers. comm.).

All researchers should be required, as a condition of the NPS research permit and expressly stated so in the Research Permit Reporting System (RPRS) form, to provide the sponsoring park the following items:

- digital or hard copies of all field notes and field photos,
- GPS positions of all collection localities including measured sections, fossil specimens, sediment samples, boreholes, and rock samples for acetate peels, thin sections, geochemical or paleomagnetic research
- an Excel spreadsheet of all specimens to be cataloged and/or cited in publications, theses, or dissertations.

Spreadsheet fields should be set up to include all information classes required by ANCS+ allowing for proper cataloging of NPS material. Templates are available through Guadalupe Mountains National Park (GUMO) if needed (G. Bell, pers. comm.); information can be directly imported in ANCS+ once spreadsheet fields are populated.

The researcher should additionally be required to place NPS catalog numbers and identifying labels on all collected specimens and ensure that all items are cataloged into the off-park repository named in the permit. Although the researcher is responsible for providing this information to the park, NPS staff could offer support in completing such items as researchers may be unfamiliar with NPS protocol.

During the RPRS application process, a researcher must obtain signatory agreement from the collections manager or curator of any off-park repository to house collections made as a consequence of the permit. This is the point at which conditions should be negotiated, if for nothing else but to inform all parties of the scope of commitment involved. If the researcher or officials of the repository will not commit to doing the required work after reasonable negotiations, a permit should be denied. If the conditions of the permit are not fulfilled in a timely manner, then the park has reasonable cause not to issue further permits to that researcher until conditions are met. This strategy is consistent with requirements in the Natural Resources Management Guidelines (NPS-77). Although other federal laws and NPS policies offer protection of government property and park resources in general, NPS-77 objectives specifically focus on the preservation of fossils for their historic and scientific value. This can only be achieved if effective communication is maintained between the park, repositories, and researchers before, during, and after fieldwork is conducted.

The park also has a role in the satisfactory completion of research, in that it should be receptive to the needs of a researcher in fulfilling the requirements of a research permit. Depending on the conditions of the permit, assistance from park staff in obtaining information, equipment, or access to localities or collections may be requested in which case it is essential that a park follow through with obligations made during the permitting process. The permit process ensures that researchers are fulfilling requirements put forth by the NPS; however, open communication between park staff and visiting scientists is a critical part of this process and that collaboration ultimately allows for the successful stewardship of park resources.

Table 2. Taxa listed in Amistad National Recreation Area paleontological collections held at park headquarters. There is no locality information associated with specimens listed below from "Existing Collections", although they are from Cretaceous units within the park. The "New Collections" were made during the senior author's locality visits in 2006. The senior author also provided a fossiliferous slab from the Late Paleozoic for interpretive purposes. This slab was found out of context in the vicinity of Diablo East.

Existing Collections	New Collections
Class Cephalopoda	<u>Salmon Peak Limestone</u>
Cymatoceras sp.	**Class Gastropoda**
Pervinquieria sp.	*Nerinea* sp.
Plesioturrilites sp.	**Class Bivalvia**
unidentified nautiloid	unidentified rudist
Class Gastropoda	<u>Salmon Peak Limestone</u>
Nerinea sp.	**Class Gastropoda**
Tylostoma sp.	*Nerinea* sp.
	Class Bivalvia
Class Bivalvia	unidentified clam
Exogyra ponderosa	
Ilymatogyra arietina	<u>Del Rio Clay</u>
Lima sp.	**Class Bivalvia**
Neithea sp.	*Ilymatogyra arietina*
Trigonia sp.	
unidentified clams	<u>Del Rio Clay</u>
unidentified rudists	**Class Bivalvia**
	Ilymatogyra arietina
Suborder Textulariidae	
Cribratina texana	<u>Buda Limestone</u>
	Class Echinoidea
	Salenia sp.
	Class Bivalvia
	unidentified rudist

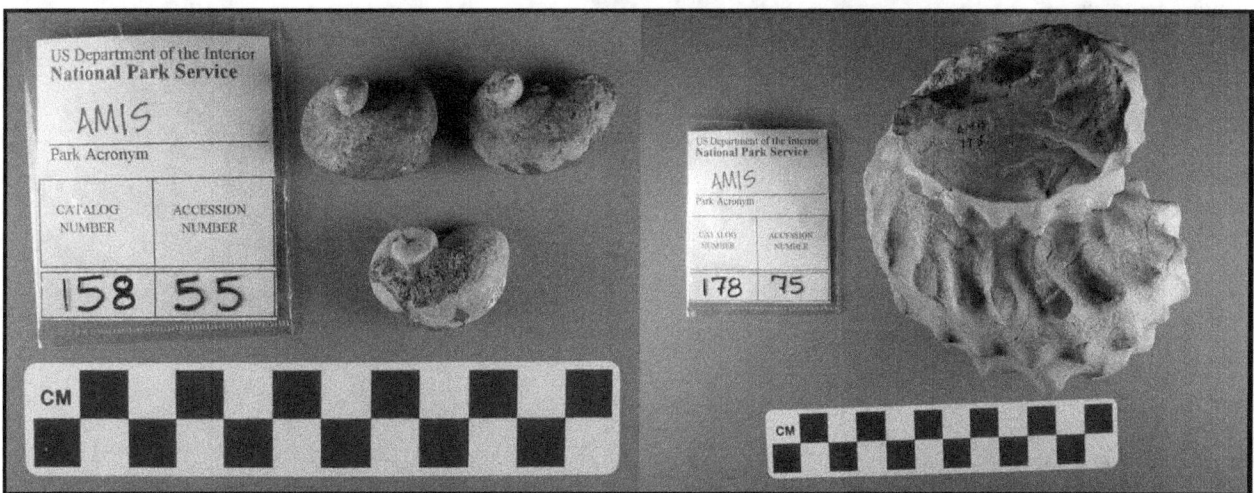

Figure 7. Sample of fossils within Amistad National Recreation Area museum collections. Left: *Ilymatogyra arietina* oysters showing signs of chemical alteration. Right: A partial ammonite, *Plesioturrilites* sp., from Cretaceous deposits. Photos by Jack Johnson.

Table 3. Taxa listed in Amistad National Recreation Area collections for Cueva Quebrada and Conejo Shelter.

Cueva Quebrada	Conejo Shelter
Class Gastropoda	**Class Actinopterygii**
unidentified snails	*Carpiodes carpio*
Class Amphibia	*Ictalurus* sp.
Rana sp.	*Ictalurus furcatus*
unidentified anurans	*Lepisosteus* sp.
Class Reptilia	*Lepisosteus osseus*
unidentified lacertilians	*Micropterus* sp.
unidentified chelonians	*Pylodictis olivaris*
unidentified squamatans	*Scaphirhynchus platorynchus*
Class Aves	unidentified castostomids
unidentified Falconiformes	unidentified ictalurid
unidentified Passeriformes	unidentified teleost
unidentified birds	**Class Reptilia**
Class Mammalia	*Trionyx* sp.
Ammospermophilus interpres	unidentified emydid
Arctodus sp.	**Class Aves**
Baiomys sp.	unidentified bird
Bison sp.	**Class Mammalia**
Camelops sp.	*Bassariscus astutus*
Canis latrans	*Lepus californicus*
Equus sp.	*Neotoma* sp.
Equus francisci	*Sylvilagus* sp.
Equus scotti	
Lepus sp.	
Mephitis sp.	
Navahoceros sp.	
Neotoma sp.	
Onychomys leucogaster	
Pappogeomys sp.	
Perognathus sp.	
Reithrodontomys sp.	
Spilogale sp.	
Stockoceros sp.	
Sylvilagus sp.	
Thomomys bottae	
Urocyon cinereoargenteus	
unidentified carnivores	
unidentified chiropterans	
unidentified cricetids	
unidentified geomyid	
unidentified heteromyid	
unidentified lagomorphs	
unidentified leporids	
unidentified rodents	
unidentified mammals	
unidentified mustelids	
unidentified ungulates	
unidentified vespertilionid	

Interpretation

Because archeological resources are the main focus of interpretive programs at AMIS, paleontology is often overlooked in interpretation despite the abundance of fossils inside park boundaries. This report attempts to provide paleontological resource information that could support future interpretive efforts. Literature concerning park fossils is limited and existing specimen collection records were mostly inadequate; consequently, paleontology is rarely mentioned in current park programs, exhibits, or publications. There are a few exceptions that shall be described below; however, much more could be offered that specifically focuses on the local geology and paleontology of AMIS.

Current Efforts

Dino Days

There is an annual paleontological program called Dino Days presented jointly by AMIS and the Whitehead Memorial Museum in Del Rio. The program began in 2002 and is geared toward school groups in kindergarten through third grade. Events at Dino Days include crafts, a main presentation, hands-on and visual activities, modern bones used to compare against large paper mache models of *Apatosaurus* bones, a global map of documented dinosaur finds, and a simulated dinosaur dig borrowed from the Texas Memorial Museum. While this program may be a wonderful opportunity to engage school children in paleontology, local marine fossils are not utilized at all and similarly dinosaurs are not preserved in the AMIS vicinity. This could present an interpretive opportunity. Why not ask the school children why no dinosaurs are found near AMIS? The area was underwater! Dinosaurs from the same age as the rocks within AMIS are found in north central Texas. The dinosaur fossils found within Big Bend National Park and the surrounding area are also Late Cretaceous in age, but are slightly younger (90 to 65.5 million years old) than the Upper Cretaceous rocks at AMIS (99 to 95 million years old).

Informal Programs

Informal geology programs that mention paleontological resources found in local limestone units are available in the form of in-class or houseboat presentations given by Lisa Evans, AMIS Educational Specialist. Programs such as these are not regularly scheduled.

Wayside Panels

There are a few panels inside the park that provide information on local geological and paleontological resources. One wayside describes an ancient sea fully of shelly fauna that eventually contributed to the formation of limestone cliffs at AMIS. Another sign describes domestic midden deposits composed of plant fragments and fish bones.

One wayside installed after paleontological fieldwork in March 2006 specifically focuses on stratigraphic units in the park and the characteristic fossils therein. This panel labeled *Under the Sea* conveys information on fossilization, fluctuating paleoenvironments, and oysters and scallops frequently unearthed from the Upper Cretaceous Del Rio Clay (Fig. 8). Interest in designing this sign is a reflection of heightened paleontological awareness of park staff following fieldwork undertaken for this report; however, increased monitoring in the vicinity of this wayside is now strongly encouraged.

Potential Interpretive Resources

Accessible Fossils

Diverse fossils are cataloged in the collections at park headquarters. These have served as a resource in previous presentations on the geology of the region. Although fossils should be utilized for interpretive purposes, setting aside specimens for an education collection provides a more accessible set of fossils reserved specifically for interpretation. These paleontological resources may not (and should not) be of high scientific value. Several Smokey Lehnart specimens that lack stratigraphic and locality information (and are represented elsewhere in the collections) are ideal candidates for the education collection.

Fossils that exhibit different forms of preservation and mineralization might be useful for thematic interpretative programs. Touch and feel specimens are particularly engaging for younger audiences. Identifying organisms based on their hard parts (shells) or behavioral remains (ichnofossils) are other interesting subjects that could be explored for interpretation. Specimens belonging to the Shumla School might serve as additional resources if paleontological programs are developed in cooperation with AMIS. Jack Johnson started the development of a paleontological program for the Shumla School based on this report, and continued cooperative efforts with the Shumla School are likely possible.

Photographs

Photographs of *in situ* and collection specimens made available through this research effort could serve as a starting point for self-guided interpretation opportunities at the visitor center. Park staff could include depictions of paleontological resources in slide or PowerPoint presentations as well as design photo albums highlighting fossils for visitors to peruse on their own, perhaps accompanied by a website for those interested in learning more after their experience at the park.

Displays

Accessible fossils and photographs provide interpretive opportunities until a more permanent exhibit can be created for AMIS paleontological resources. Fossils are not presently displayed at the park, but plans for a new visitor center should provide supplemental space for exhibits at the park.

Other resources include literature in the park library and books available for purchase at the visitor center such as geological road guides and dinosaur books for children. Field guides by Matthews (1978) and Finsley (1999) that focus on identifying fossils from Texas should be added to the park library as they could greatly aid in a variety of interpretative efforts.

Interpretive Potential

This report provides a review of fossiliferous formations and localities in the park as well as common fossils of both the Cretaceous and Quaternary in the AMIS vicinity. This information could be used in developing interpretive materials and programs to increase public paleontological awareness. The use of current waysides might be an easy way to start exploring the geologic history of the region. The addition of paleontological pamphlets or scheduling guided hikes could further strengthen interest in these non-renewable resources at AMIS. Topics for specialized programs might include:

- characterization of fossils vs. pseudofossils (inorganic features found in rocks that superficially resemble fossils)

- modes of fossilization

- history of life and evolution

- paleoenvironmental reconstruction of southwest Texas

- significance of fossils

- archeology vs. paleontology

- fossiliferous fire-cracked rock

- paleontological misconceptions

Utilizing fossils in reconstructing ancient environments might be especially interesting being as such a drastic change is observed at AMIS from the marine realm of the Cretaceous to the woodlands and deserts of the Pleistocene and Holocene. This also presents an opportunity to discuss climate change.

School groups and larger local audiences might enjoy programs that help them identify rocks and fossils in their own backyard. Matching cards, handling specimens, coloring pages or crayon rubbings of common fossil outlines and their scientific names, as well as comparisons to modern flora and fauna might be appealing activities for families staying at the park. Scientific lectures or guided field trips led by visiting researchers or through the Shumla School, Seminole Canyon State Park and Historic Site, or Sul Ross State University offer other educational opportunities. Discussing the importance of resource protection and the NPS stewardship mission should be an important component of all paleontology interpretation.

The prospect of a new visitor center could additionally further interpretive programs allowing more space for exhibits. One display that might be particularly informative and interesting is a geological time line demonstrating when various fossils fit into the history of life such as AMIS marine shells from the Cretaceous and bison jumps from the much younger Holocene. Other favorites include Paleozoic trilobites, Mesozoic dinosaurs, and characteristic or official

state fossils of Texas. For example, the official state dinosaur of Texas is the Early Cretaceous long-necked *Pleurocoelus* found in rocks older than those at AMIS. The Texas state stone is petrified palm wood much younger (~30 million years old) than the Mesozoic fossils at AMIS.

Interpretive Preparation

The interpretive staff welcomes suggestions for future paleontological programs. This should begin by ensuring that staff are adequately instructed on various aspects of AMIS geology and paleontology before attempting to relay information to the public. Descriptions of common AMIS fossils, stratigraphic units, and nearby localities ideal for familiarizing staff with the geology of the region are provided in this report. This information could be reviewed and selectively used for paleontological programs.

Interpretation and Resource Management

Specifics regarding sensitive localities are not to be disclosed to the public as increased risk could be placed on paleontological resources requiring additional protection and monitoring. It might be desirable to direct public attention to areas that may not be significantly endangered by increased visitation. This has been suggested for other parks in the form of a paleontological disclosure policy in that localities are ranked allowing for particularly high risk or unique deposits to be protected. Park areas open to the public can then be utilized to enlighten visitors about paleontological resources and the significance of *in situ* fossils and the NPS stewardship mission, instilling a desire to protect fossils as opposed to pocketing them for personal use or profit.

The public should be informed that fines are applied if fossils are removed from federal lands and it is helpful also to remind visitors that fossils can only be enjoyed by everyone if they remain in the park. Training interpretive staff should additionally address paleontological issues concerning public misconceptions, media misinformation, and controversial opinions surrounding evolution and geologic time. Techniques for exploring such topics have been considered at previous NPS Fossil Resource Conferences.

Increasing public awareness of fossils in the park could increase potential impacts to AMIS paleontological resources. A paleontological resource management plan (including monitoring strategies) may help address these concerns and provide an avenue to identify increased impacts or threats. Most fossils are inaccessible or firmly cemented in local rocks alleviating some concerns of law enforcement. Nonetheless, communication between interpretation, law enforcement, resource management, and maintenance staff could provide a foundation for park awareness of fossils, their interpretation, and public accessibility. The *Management and Protection* chapter offers some additional recommendations in this regard.

Summary

Overall, recommended actions for park interpretation could include the incorporation of additional specimens for the education collection, program development including aspects of paleontological resource stewardship, increased focus on local geological and paleontological resources perhaps through utilization of neighboring institutions, adherence to a paleontological disclosure policy during paleontological resource interpretation, and increased communication among park staff regarding potential impacts of paleontological resource interpretation.

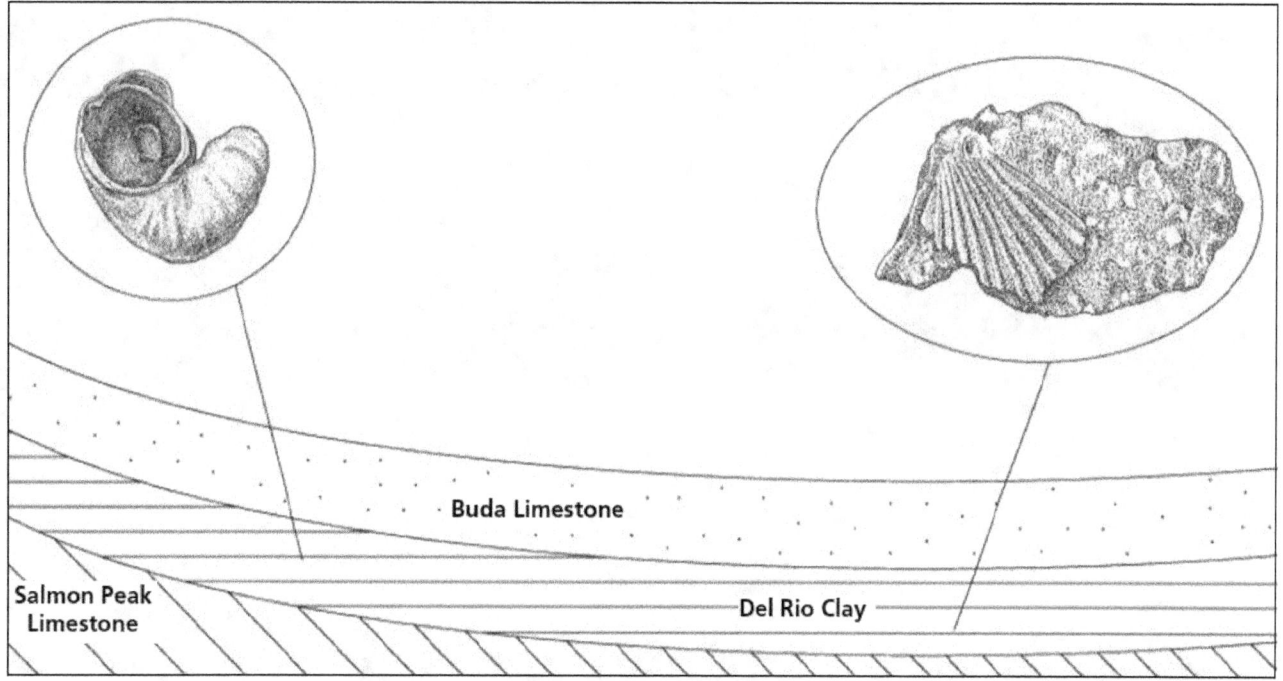

Figure 8. Image modified from the "Under the Sea" wayside created for the park. Left: An *Ilymatogyra arietina* oyster characteristic of the Del Rio Clay. Right: A pectinid scallop accompanied by Del Rio Clay oysters. Artwork by Jack Johnson based on specimens from AMIS collections.

Management and Protection

Management plans at Amistad National Recreation Area generally do not address protection of paleontological resources. Bones, shells, and other specimens collected during archeological excavations are often incorporated; however, marine fossils from the Mesozoic are left unmentioned. The need for research and inventory of AMIS paleontological resources is briefly stated in the Natural Resources Management Plan (National Park Service 1974), yet no formal research regarding fossils in the park commenced until this investigation in March 2006.

Limited paleontological resource management can be partially attributed to the absence of information regarding AMIS geology and fossils prior to this report. The only report that presently refers to the care of both Cretaceous and Quaternary specimens is the Collection Management Plan (Labadie et al. 2005). This document is summarized in a previous chapter; however, protection of *in situ* fossils and paleontological localities likely requires further attention.

The NPS held its Geologic Resources Inventory (GRI) scoping meeting at AMIS in April 2008 (KellerLynn 2008). The original draft of this manuscript served as a starting point for information regarding the geologic formations exposed in the park and the fossils therein. As a result of discussions on paleontological resources, Greg Garetz, Chief of Education and Resource Management, offered accounts of attempted fossil theft at the park, further described in the "Threats" section of this chapter. Protection of NPS fossils is now aided by the Paleontological Resource Preservation Act, which should facilitate the development of management plans for NPS paleontological resources nationwide.

Policies and Data Supporting Paleontological Resource Management

The Paleontological Resource Preservation Act (PRPA) (P.L. 111-11), signed into law on March 30, 2009, will serve as explicit authority for the management, protection and interpretation of paleontological resources in parks, in addition to the NPS Organic Act of 1916. The PRPA specifically provides the NPS with the following mandates to enhance paleontological resources stewardship:

- §6302 calls for the management and protection of paleontological resources using scientific principles and expertise. The NPS should develop plans for inventory, monitoring, and the scientific and educational use of paleontological resources. Planning should emphasize interagency coordination and collaboration and where possible include non-federal partners, the scientific community and the public.

- §6303 calls for the establishment of education programs to increase public awareness about the significance of paleontological resources.

- §6304 calls for the development of a specific permit for the collection of NPS paleontological resources. The new legislation and other existing authorities clarify and reaffirm issues of property ownership, accountability, access and confidentiality of locality information associated with the management of NPS paleontological resources.

- §6305 calls for the curation of NPS paleontological resources, along with any associated data or records, in approved repositories.

- §6306 provides clarity regarding prohibited acts involving paleontological resources and specifies criminal penalties associated with these prohibited acts.

- §6307, along with other existing authorities, enables the NPS to seek civil penalties and restitution for the violation of any prohibited activities involving paleontological resources.

- §6308 provides the NPS a confidentiality provision with an exemption from the disclosure of any information associated with the nature and specific location of NPS paleontological resources.

- §6310 directs the Secretaries of the Interior and Agriculture to issue regulations appropriate to carry out the Act.

In addition to the PRPA, several other authorities influence the management of NPS paleontological resources. The 2006 National Park Service Management Policies (§1.4.6) stipulates that paleontological resources are considered park resources and values that are subject to the "no impairment" standard set forth by the NPS Organic Act in 1916. Basic guidelines for management of paleontological resources are found in sections 4.8.2 and 4.8.2.1 of the 2006 NPS Management Policies.

NPS paleontological resource management is also guided by Natural Resources Management Guidelines (NPS-77). Objectives include:

- identification of NPS paleontological resources

- evaluation of their significance

- protection in order to preserve their historic and scientific value

- support of management goals through research

The first step in creating a paleontological resource management plan is to research the resource needing protection. The presence of fossils on federal land is reported in early surveys of archeological resources by Graham and Davis (1958). They commented that salvage considerations should be similarly provided for paleontological resources prior to reservoir completion; however, only Quaternary specimens at archeological localities benefited from this suggestion. The need for a paleontological survey focused on Cretaceous resources at AMIS was overdue considering the extent of fossiliferous exposures in the region. Anderson (1974) noted a "lack of geological and paleoenvironmental research for the Amistad area" in his report on archeological localities. The Natural Resources Management Plan for Amistad NRA (National Park Service 1974) similarly recommends documentation of paleontological localities - goals addressed during this investigation. This report provides a baseline level of information for four characteristics required for documenting fossil localities: geographic, stratigraphic, paleontologic, and geologic data.

Geographic Data

The first component includes geographic or geospatial coverage of localities, UTM coordinates obtained using differentially corrected GPS measurements, and maps. Several maps of paleontological localities and coordinates for areas registered as "significant" resulted from this project; however, future projects could include:

- compiling a database of AMIS paleontological localities
- establishing a system for cross-referencing existing names and numbers for park areas (particularly localities of shared archeological significance)
- obtaining GPS data for all areas containing paleontological resources (not possible in March 2006)

Stratigraphic Data

Stratigraphic units and corresponding ages comprise the second requirement of paleontological surveys. The range of stratigraphic units exposed in the park and their relative ages are provided in earlier chapters of this report.

- Further investigation is needed to confirm the presence and/or extent of the Boquillas Formation, Eagle Ford Group, and Austin Chalk inside park boundaries
- High resolution digital geologic maps for AMIS should be available late in 2009 as a result of NPS Geologic Resource Inventory efforts and can aid in investigating park exposures

Paleontologic Data

Identification of fossil taxa within the park encompasses paleontological information. This survey did not focus on species-level classification; however, genus names are provided for many specimens in the collections at park headquarters. Identification of *in situ* fossils was limited.

- Species-level identification could provide an additional level of detail for continued and/or future research
- Enhanced documentation of *in situ* specimens is proposed

Geologic Data

The final component includes descriptions of lithology and geological interpretations regarding depositional environments of fossiliferous layers. Field observations and a review of the literature proved useful in uncovering this material.

Future Paleontological Research Opportunities

While a wealth of paleontological information is provided by this investigation, opportunities for additional research and field surveys are plentiful. Inventorying paleontological resources for Cretaceous and Quaternary localities throughout the park could continue based on the information in this report. Presentation and publication of this report

may serve as a catalyst for future research. There are no paleontological research permits currently active at AMIS, but subjects for future exploration include:

- archeological localities that may contain fossil remains
- inundated outcrops requiring examination during lower lake levels
- fossiliferous sites that partly fall upon private land
- exposures difficult to access along the river west of Langtry
- fossiliferous units that extend across the reservoir into Mexico

Anderson (1974) conducted a preliminary archeological survey of 68 sites in the vicinity of AMIS. Although he did not differentiate AMIS areas from localities found outside park boundaries, name, number, type of site (e.g., rock shelter, open midden, etc.), associated resources (e.g. pictographs, perishable remains, etc.), and summarized references are listed for many localities. His guide may serve as an excellent starting point in seeking fossils at archeological sites. Dering (2002) is another superb resource. He offers a cultural chronology, classification of federal vs. other jurisdictional areas, information on shoreline and regional surveys, database recommendations, impacts affecting cultural resources (many of these hazards similarly affect fossils), and optimal areas for park expansion opportunities.

Likewise, geology publications prior to the filling of the reservoir in the late 1960s including IBWC reports should be useful in finding fossiliferous exposures that may be presently inundated. Episodes of lower water levels at AMIS should be utilized not only for revisiting archeological localities, but similarly for surveying paleontological resources. Photo documentation and specimen excavation could be completed for sites that are otherwise inaccessible under normal lake level conditions. Exposures beyond Langtry remained unexplored due to accessibility issues in this investigation; nonetheless, areas must be examined for potential paleontological resources. Lastly, localities that fall in part upon private land and outcrops that cross the border into Mexico will require cooperation from local land owners along the Devils River, Pecos River, and Rio Grande.

Threats

Once fossiliferous localities are documented, threats to paleontological resources at those localities can be identified including erosion, unstable strata, excessive visitor use, fossil theft, development, vandalism, and more.

Anderson (1974) recommended the study of clay weathering and rock fall frequency at Bonfire Shelter as such events can impact archeological resources. Fossils in southwestern deserts may also be severely affected by erosional processes. Loss of rock can result from rain, stream runoff, wind, mineral growth, intrusive vegetation, and animal disturbance (e.g. Santucci et al. in press). While erosion can be beneficial in unearthing new paleontological resources, repeated loss of sediment will eventually cause damage to fossiliferous exposures at the surface and below ground. Softer sediments of the Del Rio Clay are particularly prone to erosion.

Increased access to localities following the rise of the reservoir in the late 1960s is further noted by Anderson (1974) as a heightened concern for potential poaching and vandalism. Management of these sites is complicated by the fact that they are often situated above the elevation boundary for the park; however, NPS localities along the reservoir are commonly protected by fencing or regular patrols. It is important to preserve non-renewable resources found above normal pool level as most of those influenced by flooding are destroyed (Anderson 1974). Although leaving specimens *in situ* is the preferred management strategy of the NPS, excavating rare or exceptionally well-preserved fossils could be undertaken as needed, depending on whether such paleontological resources are in immediate danger of anthropogenic or natural hazards.

Many of the comments by Anderson refer to archeological localities; however, similar human-induced threats affect paleontological resources. Although scientific research permits are required to retrieve fossils from inside the park, illegal collecting for both personal and commercial use can be problematic as it has been in the past at the park for both archeological and paleontological resources.

Management actions to alleviate illegal collection of paleontological resources and minimize threats by artificially accelerated erosion are supported by management policies of the National Park Service and the Paleontological Resource Preservation Act of 2009. Measures of prevention include construction of shelters over *in situ* specimens, stabilization of unsteady exposures, or removal of fossils to museum collections. Further suggestions of monitoring, prospecting, reburial, park restrictions, and regular patrols are listed in the Natural Resources Management Guidelines (NPS-77).

As part of a forthcoming (late 2009) geologic resource monitoring manual, Santucci et al. (in press) present a chapter outlining suggestions for monitoring paleontological resources *in situ*. The chapter includes specific discussion of paleontological resources found along shoreline features and includes case studies from another reservoir recreation area, Curecanti National Recreation Area in Colorado. This chapter may be useful for developing monitoring strategies at AMIS.

As further described in Santucci et al. (in press), periodic reassessment and monitoring of localities should be accompanied by photo documentation. Prospecting in areas subject to lower water levels or heightened erosion may be useful in uncovering new resources. Threatened specimens may be reburied if unable to be salvaged in the immediate future. Temporarily restricting public access in areas requiring prolonged investigation limits public knowledge of significant paleontological exposures. Organizing regular patrols at important localities may reduce the occurrence of poaching and vandalism. The AMIS Natural Resources Management Plan (National Park Service 1974) further recommends visitor use analyses as such information can be utilized in recognizing specific risks for paleontological localities. Detailed social science studies such as those completed at Petrified Forest National Park and Fossil Butte National Monument can provide additional information regarding visitor attitudes toward paleontological resources.

Theft of paleontological resources is not commonly reported at AMIS; however, it has been observed. G. Garetz (pers. comm.) has intercepted individuals on several occasions who were attempting to load rocks into their vehicles to be used for decorative landscaping at their homes. Most instances involved theft of non-fossiliferous limestone, but one couple had attempted to collect fossil-bearing rocks in the park. The husband and wife had claimed that they were merely preserving fossils which would eventually be lost after being submerged again upon a rise in lake level. Water offers some protection for paleontological resources from collection, but presents a variety of other impacts to fossil resources (Santucci et al. in press). A drop in lake level increases their exposure and leaves fossils more vulnerable to illegal collection. Examples described by G. Garetz suggest that fossil theft is a reality at AMIS and such violations should be addressed in management plans specific to paleontological resources.

Recommended Actions

The preservation of fossils is important for scientific research, interpretation, and public enjoyment. Research on paleontological resources by the academic community is encouraged. However, NPS parks need to enhance knowledge of their own resources through inventory and monitoring programs. Seeking collaborative partnerships in government agencies, academic institutions, and public and private organizations may help in conserving paleontological resources.

Fossils are not the focus of current management plans at AMIS. Although most visitors recognize that removing natural and cultural objects from national parks is illegal, no specific form of protection was available for paleontological resources until the very recent (March 2009) passage of the Paleontological Resources Preservation Act. The Superintendent's Compendium (2006) states that collection of natural specimens from the park is not permitted, but specific mention of fossils and paleontological localities is absent from that document. This is a concern as small specimens could be pocketed easily from inside park boundaries and sold by local dealers in rock shops and souvenir stores (or used for decorative landscaping as mentioned earlier). Fossils may benefit from restricted public access in rock shelters (at least until the development of a Cave and Karst Management Plan); however, many paleontological resources in ground, road, or cliff outcrops are not covered by regulations regarding rock shelters. Protection of archeological localities may in part aid fossils found in a cultural context. The NPS Cultural Resources Management Guidelines (NPS-28) and Archeological Resources Protection Act of 1979 offer guidance regarding paleontological specimens found in archeological settings; nonetheless, management plans should be created specifically for paleontological resources as natural non-renewable resources commonly require different measures of protection. Other NPS regulations protecting archeological localities such as restriction of campgrounds and campfires should be extended to cover park paleontological localities in the Superintendent's Compendium (2006).

The development of a paleontological resources management plan is necessary to ensure protection of *in situ* fossils at

AMIS. Options may include patrol of paleontological localities (particularly during periods in which lake level has decreased), cooperation of neighboring organizations and land owners, limiting access at endangered localities, reducing the effects of erosion, mitigating effects of erosion by cyclic prospecting and collecting of new paleontological resources, installing alarm systems and barriers, monitoring visitor impact, recording incidents of theft and vandalism, reburying fossils at risk from human-induced hazards, and removing specimens of scientific importance. Management plans for *in situ* fossils and paleontological localities must additionally consider aspects of collection and interpretation. Several recommendations are offered in previous chapters on those subjects, but it is important to recognize that an integration of management practices is needed. The Collection Management Plan (Labadie et al. 2005) already incorporates paleontological specimens and further suggestions for archiving material and organizing fossils shall assist in future paleontological research at the park. Decisions regarding excavation of exposed specimens require the cooperation of staff in collections and natural resources; thus, management plans should consider both roles in preserving paleontological resources. Interpretation is strongly linked to management practices as heightened protection and staff preparation may be necessary following increased public awareness of park fossils. Management plans should incorporate recommendations mentioned in the *Interpretation* chapter such as adherence to a paleontological disclosure policy, offering an increased focus on stewardship for visitors, limiting public access to park localities, paleontological resource education for park staff, and choosing interpretive programs that minimally impact paleontological resources. Achieving a balance between the interpretation and protection of AMIS fossils is critical.

It is imperative that all park employees remain informed and educated regarding the paleontological resources that they are responsible for protecting. Maintenance staff should be able to recognize and understand the significance of park fossils and know the location of paleontological exposures so proper avoidance procedures can be practiced if ground-disturbing work is needed. It is also essential that law enforcement know where localities are inside the park and how to recognize fossils so they can schedule regular patrols of such areas, interdict illegal collecting, and accurately identify illegally collected resources.

An important aspect for protecting paleontological resources is managing NPS localities recognized as "significant." Having important localities officially documented and accompanied by completed condition assessments additionally helps fulfill the Government and Performance Results Act (GPRA) goal #1a9A. There are six areas at AMIS that are candidates as significant GPRA localities. A review of such paleontological resource issues may facilitate the development of a management plan. Some localities contain exceptional or endangered paleontological resources; others are candidates for interpretation or already receive protection from existing management programs in archeology. Although efforts are made by law enforcement to stop the illegal collection of rocks throughout the park, informing AMIS staff of particularly important localities raises park awareness of paleontological resources requiring special attention. All areas should be periodically patrolled and the frequency of patrols based on accessibility and visitor impact.

Preservation of NPS paleontological resources is essential as such fossils contribute to an understanding of the history of life on earth. Non-renewable resources at AMIS require special protection as reviewed in Anderson (1974) and as stated by Gingrich in the previous year, "The structure of the limestone is soft, and the shelters are constantly exposed to weathering. An even greater menace is man himself."

Summary of Recommendations

Amistad National Recreation Area has a rich array of Cretaceous and Quaternary paleontological resources. This survey aims to raise paleontological awareness at the park by providing a review of park geology, paleontological localities, and collection updates. Future interpretative programs, research opportunities, and management options are additionally considered. This section comprises a summary of the main recommendations for AMIS. Additional recommendations are also found in the appropriate sections of the report.

Paleontological resources are a promising source of scientific and public interest at the park. This report provided a summary of baseline information regarding various aspects of AMIS paleontological resources and offered recommendations for their future management Because much of the park is underlain by fossiliferous material, an understanding of paleontological resources within the park provides a strong foundation for the future of natural resource stewardship at Amistad National Recreation Area.

Museum Collections and Curation

- Further research is needed to ascertain whether localities mentioned in the literature fall within NPS boundaries. Specimens collected inside the park during such research efforts should be moved to park repositories.

- The small collection of Cretaceous specimens at TARL could be removed to park headquarters.

- Enhanced inspection of Quaternary fauna and flora found in archeological collections is recommended for a more complete inventory of paleontological resources.

- Taxonomic names should be clarified (and if possible identified to species level) for park fossils particularly for large and unique rudistid bivalves.

- AMIS collections could further benefit from an increased focus on archives, images, and educational specimens, all potentially useful in creating interpretive fossil programs for park visitors.

- Templates for entering data into ANCS+ are available from GUMO and a PMIS statement should be submitted to acquire funds for any backlog cataloging needed at AMIS.

Interpretation

- The abundance of Cretaceous and Quaternary resources at AMIS could be increasingly utilized for interpretive programs; however, a loan system is needed if specimens are to be used from park collections. Compiling a separate collection of specimens for interpretation is recommended.

- Development of cooperative programs may be possible through the Shumla School, Seminole Canyon State Park and Historic Site, and perhaps neighboring academic institutions. Existing programs such as Dino Days could incorporate more information on local marine fossils.

- Plans for a new visitor center could incorporate space for exhibits on paleontological resources of the region.

- Existing panels, waysides, and nearby localities not easily endangered by increased visitation could be utilized more for paleontological programming.

- A stewardship message for visitors and a paleontological disclosure policy for safeguarding fossiliferous localities should be included in all interpretive efforts.

Resource Management and Protection

- Ongoing research of fossiliferous localities and park specimens is encouraged. New exposures of species-rich deposits and unexplored areas or units including the Boquillas Formation may be discovered inside park boundaries.

- Important fossil localities, particularly fossil-rich localities, and rapidly eroding localities should be prospected and collected on a regular basis.

- Training all park staff in local geology and paleontology, particularly those working in the field is suggested.

- The six significant localities identified within the park as a part of this investigation provide a foundation for management and protection of *in situ* paleontological resources.

- A paleontological resource management plan could be helpful in identifying threats and, as appropriate, suggest actions such as instituting patrols, restricting access, building physical barriers, and salvaging important specimens as needed.

- Incorporating specific mention of paleontological resources in the Superintendent's Compendium and the developing Cave and Karst Management Plan could aid in protecting park fossils until a dedicated paleontological resource management plan is created.

- Discussions of paleontological resources found along shoreline features for another NPS reservoir in Colorado, Curecanti National Recreation Area, as presented by Santucci et al. (in press), may be useful for developing monitoring strategies for AMIS paleontological resources.

- Ongoing locality assessments through periodic monitoring, prospecting, and protecting inundated exposures during lower lake levels, and obtaining permission from local land owners to manage sites that partly fall upon private land is recommended.

- Future management efforts could benefit from the development of a park database of paleontological localities that incorporates GPS information and management notes from related archeological reports.

- Future addition of high resolution geologic maps as part of ongoing the NPS Geologic Resource Inventory efforts should aid in this process.

Literature Cited

Aepli, D. B. 1989. Biostratigraphy and paleoecology of the Grayson Formation, McLennan County, central Texas. Bachelor's Thesis. Baylor University, Waco, TX.

Albritton, C. C., Jr., J. R. Puryear, and W. W. Schell. 1954. Foraminiferal populations in the Grayson marl [Texas]. Geological Society of America Bulletin 65(4):327-336.

Alexander, R. K. 1974. The archeology of Conejo Shelter: A study of cultural stability at an archaic rockshelter site in southwestern Texas. PhD Dissertation. University of Texas, Austin, TX.

Alexander, R. K. 1970. Archeological investigations at Parida Cave, Val Verde County, Texas. Papers of the Texas Archeological Salvage Project No. 19, Austin, TX.

Alvarado-Ortega, J., A. Blanco-Pinon, and H. Porras Muzquiz. 2006. Primer registro de Saurodon (Teleostei, Ichthyodectiformes) en la cantera La Mula, formacion Eagle Ford (Cretacico superior; Turoniano), Muzquiz, Estado de Coahuila, Mexico. Revista Mexicana de Ciencias Geologicas 23(1):107-112.

Anderson, B. A. 1974. An archeological assessment of Amistad Recreation Area, Texas. National Park Service Division of Archeology, Southwest Region, Santa Fe, NM.

Archer, K. 1936. Some Cephalopoda from the Buda Limestone. Master's Thesis. University of Texas, Austin, TX.

Ashmore, R. A., J. B. Stevens, D. A. Reed, and M. S. Stevens. 1997. Possible-different uses of giant inoceramid clustering, middle to upper Boquillas Formation, Big Bend region, Trans-Pecos, Texas. Geological Society of America Abstracts with Programs 29(2):2.

Bardack, D. 1965. New Upper Cretaceous teleost fish from Texas. University of Kansas, Lawrence. Kansas University Paleontological Contributions Paper 1.

Bardack, D. 1968. *Belonostomus sp.*, the first holostean from the Austin Chalk (Cretaceous) of Texas. Journal of Paleontology 42(5):1307-1309.

Barnes, V. E. 1977. Del Rio Sheet. Geologic Atlas of Texas. University of Texas Bureau of Economic Geology, Austin, TX. 1:250,000 scale map.

Barnes, V. E. 1992. Geologic map of Texas. University of Texas Bureau of Economic Geology, Austin, TX. 1:500,000 scale map.

Barrett, M. L. and J. P. Goodson, Jr. 2006. High-resolution foraminiferal biostratigraphy of Cenomanian and Turonian sandstones, Tyler County, Texas. Transactions of the Gulf Coast Association of Geological Societies 56:27-37.

Bell, B. A., P. A. Murray, and L. W. Osten. 1982. *Coniasaurus* Owen 1850, from North America. Journal of Paleontology 56(2):520-524.

Bell, G. L., Jr. 2002. The questionable monophyly of Mosasauridae. Abstract. Journal of Vertebrate Paleontology 22(3 Suppl.):35A.

Bell, G. L., Jr. and M. J. Polcyn. 1995. A new basal mosasauroid from the Arcadia Park Member of the Eagle Ford Shale (late middle Turonian) near Dallas, Texas. Abstract. Journal of Vertebrate Paleontology 15(3 Suppl.):18A.

Bement, L. C. 1986. Excavation of the late Pleistocene deposits of Bonfire Shelter, 41VV218, Val Verde County, Texas, 1983-1984. Texas Archeological Survey Series, University of Texas, Austin, TX.

Bilelo, M. A. M. 1969. The fossil shark genus *Squalicorax* in north-central Texas. The Texas Journal of Science 20(4):339-348.

Bishop, G. A. 1989. Cretaceous-Tertiary decapod faunas of the Southwest. Geological Society of America Abstracts with Programs 21(1):3-4.

Boese, E. 1919. On a new *Exogyra* from the Del Rio Clay and some observations on the evolution of *Exogyra* in the Texas Cretaceous. University of Texas, Austin. Bulletin 1902.

Bostik, W. C. 1960. Micropaleontology of the upper Eagle Ford and lower Austin groups, Big Bend National Park, Texas. Master's Thesis. Texas Tech University, Lubbock, TX.

Bowman, T. E. 1971. *Palaega lamnae*, new species (Crustacea; Isopoda) from the upper Cretaceous of Texas. Journal of Paleontology 45(3):540-541.

Brown, C. W. and R. L. Pierce. 1962. Palynologic correlations in Cretaceous Eagle Ford Group, northeast Texas. Bulletin of the American Association of Petroleum Geologists 41(12):2133-2147.

Bryant, V. B., Jr. 1968. Late Quaternary vegetation and climate in the Amistad Reservoir area of southwest Texas. Pages 366-367 *in* Geological Society of America, Boulder, CO. Special Paper 115.

Bryant, V. B., Jr. 1977. Late Quaternary pollen records from the east-central periphery of the Chihuahuan Desert. Pages 3-21 *in* R. H. Wauer and D. H. Riskind, editors. Transactions of the Symposium on the biological resources of the Chihuahuan Desert region, United States and Mexico. U.S. National Park Service, Washington, DC. Transactions and Proceedings Series 3.

Bryant, V. B., Jr., and D. A. Larson. 1968. Pollen analysis of the Devil's Mouth site, Val Verde County, TX. Pages 57-70 *in* W. M. Sorrow, editor. The Devil's Mouth Site: The Third Season - 1967. Papers of the Texas Archeological Salvage Project No. 14.

Bullard, F. J. 1951. Microfauna of the Del Rio Formation (Lower Cretaceous) of central Texas. Master's Thesis. University of Texas, Austin, TX.

Cheetham, A. H., J. Sanner, P. D. Taylor, A. N. Ostrovsky. 2006. Morphological differentiation of avicularia and the proliferation of species in Mid-Cretaceous *Wilbertopora* Cheetham, 1954 (Bryozoa, Cheilostomata). Journal of Paleontology 80(1):49-71.

Christopher, R. A. 1982. The occurrence of the *Complexiopollis-Atlantopollis* Zone (palynomorphs) in the Eagle Ford Group (Upper Cretaceous) of Texas. Journal of Paleontology 52(2):525-541.

Cicimurri, D. J. and G. L. Bell, Jr. 1996. Vertebrate fauna of the Boquillas Formation of Brewster County, Texas; a preliminary report. Abstract. Journal of Vertebrate Paleontology 16(3 Suppl.):28.

Clark, D. L. 1959. Texas Cretaceous ophiuroids. Journal of Paleontology 33(6):1126-1127.

Clark, D. L. 1960. *Parapuzosia* in the north Texas Cretaceous. Journal of Paleontology 34(2):233-236.

Clark, D. L. and K. J. Bird. 1966. Foraminifera and paleoecology of the upper Austin and lower Taylor (Cretaceous) strata in north Texas. Journal of Paleontology 40(2):315-327.

Cobban, W. A. 1988. *Tarrantoceras* Stephenson and related ammonoid genera from Cenomanian (Upper Cretaceous) rocks in Texas and the Western Interior of the United States. U.S. Geological Survey, Reston, VA. Professional Paper 1473.

Cobban, W. A. and S. C. Hook. 1980. Occurrence of *Ostrea beloiti* Logan in Cenomanian rocks of Trans-Pecos Texas; southeastern New Mexico and West Texas. Pages 169-172 *in* P. W. Dickerson, J. M. Hoffer, and J. F. Callender, editors. New Mexico Geological Society, Socorro. Guidebook 31.

Coogan, A. H. 1973. New rudists from the Albian and Cenomanian of Mexico and south Texas. Revista del Instituto Mexicano del Petroleo 5(2):51-83.

Crane, M. J. 1965. Upper Cretaceous ostracodes of the Gulf Coast area. Micropaleontology 11(2):191-254.

Dean, G. W. 2006. The science of coprolite analysis; the view from Hinds Cave. Palaeogeography, Palaeoclimatology, Palaeoecology 237(1):67-79

Dering, J. P. 1979. Pollen and plant macrofossil vegetation record recovered from Hinds Cave, Val Verde County, Texas. Master's Thesis. Texas A&M University, College Station, TX.

Dering, J. P. 2002. Amistad National Recreation Area: Archeological Survey and Cultural Resources Inventory. Intermountain Cultural Resources Management Anthropology Projects. Professional Paper 68.

Dibble, D. S. and D. Lorrain. 1968. Bonfire Shelter: a stratified bison kill site, Val Verde County, Texas. Texas Memorial Museum, University of Texas, Austin, TX. Miscellaneous Paper 1.

Douglas, C. L. 1970. Excavations at Baker Cave, Val Verde County, Texas; part 2, Analysis of the faunal remains. Pages 111-151 *in* Texas Memorial Museum, Austin, TX. Bulletin 16.

Fielitz, C. and B. Cornett. 2002. A new elopomorph fish from the Austin Chalk Formation of Fannin County, Texas. Abstract. Journal of Vertebrate Paleontology 22(3 Suppl.):23.

Fielitz, C. and D. Bardack. 1992. *Deltaichthys albuloides*, a new and unusually preserved albulid (Teleostei). Journal of Vertebrate Paleontology 12(2):133-141.

Filkorn, H. F., C. C. Johnson, A. Molineux, and R. W. Scott, editors. 2005. Seventh International Congress on Rudists; Abstracts and Post-Congress Field Guide, Austin, TX. SEPM Miscellaneous Publication 6.

Finsley, C. E. 1999. A field guide to fossils of Texas. Gulf Publishing. Lanham, MD.

Freeman, V. L. 1964a. Geologic map of the Langtry Quadrangle Val Verde County, Texas. U.S. Geological Survey, Reston, VA. Miscellaneous Geologic Investigations Map I-422. 1:62,500 scale map.

Freeman, V. L. 1964b. Geologic map of the Shumla Quadrangle, Val Verde County, Texas. U.S. Geological Survey, Reston, VA. Miscellaneous Geologic Investigations Map I-424. 1:62,500 scale map.

Freeman, V. L. R. 1961. Contact of Boquillas Flags and Austin Chalk in Val Verde and Terrell counties, Texas. Bulletin of the American Association of Petroleum Geologists 45(1):105-107.

Friedman, V. 2001a. A new Upper Cretaceous (Cenomanian) fossiliferous locality in north central Texas. Geological Society of America Abstracts with Programs 33(5):54.

Friedman, V. 2001b. A new Upper Cretaceous (Cenomanian) fossiliferous locality in north central Texas. Proceedings of the North American Paleontological Convention 21(2):55.

Friedman, V. 2004. Paleontology, paleoecology and depositional environment of the lower Eagle Ford Group in north central Texas. Master's Thesis. University of Texas at Dallas, Richardson, TX.

Friedman, V. and A. P. Hunt. 2004. Fossil pearls from the Upper Cretaceous of Texas. Geological Society of America Abstracts with Programs **36**(5):62.

Frush, M. P. and D. L. Eicher. 1975. Cenomanian and Turonian foraminifera and palaeoenvironments in the Big Bend region of Texas and Mexico. Pages 277-301 *in* Geological Association of Canada, St. John's, Newfoundland. Special Paper 13.

Gale, A. S., J. M. Hancock, W. J. Kennedy, M. R. Petrizzo, J. A. Lees, I. Walaszczyk, and D. S. Wray. 2008. An integrated study (geochemistry, stable oxygen and carbon isotopes, nannofossils, planktonic foraminifera, inoceramid bivalves, ammonites and crinoids) of the Waxahachie Dam Spillway section, north Texas; a possible boundary stratotype for the base of the Campanian stage. Cretaceous Research **29**(1):131-167.

Gilette, D. D. 1972. Coelomic cavity casts of Upper Cretaceous fishes in Texas. Journal of Paleontology **46**(1):50-54.

Gimbrede, L. de A. 1962. Evolution of the Cretaceous foraminifer *Kyphopyxa christneri* (Carsey). Journal of Paleontology **36**(5):1121-1123.

Graham, J. A. and W. A. Davis. 1958. Appraisal of the Archeological Resources of Diablo Reservoir, Val Verde County, Texas. Pages 82-86 *in* Archeological Salvage Program Field Office, National Park Service, Austin, Texas.

Graham, J. M. 1995. Lithostratigraphy, microfacies and foraminiferal biostratigraphy of the Santonian-lower Campanian strata in the Trans-Pecos region, West Texas. PhD Dissertation. University of Texas at Dallas, Richardson, TX.

Graham, J. M. 1997. Planktonic foraminiferal biostratigraphy and depositional setting of the Santonian-lower Campanian (Upper Cretaceous) of the Trans-Pecos region, West Texas. Geological Society of America Abstracts with Programs **29**(6):53.

Grande, T. and D. Bardack. 1996. A new elopomorph from the Eagle Ford Formation near Arlington, Texas. Abstract. Journal of Vertebrate Paleontology **16**(3 Suppl.):39.

Grice, C. R. 1948. *Manorella*, a new genus of foraminifera from Austin Chalk of Texas. Journal of Paleontology **22**(2):222-224.

Hamm, S. A. 2003. Ptychodontid sharks in the Upper Cretaceous Eagleford Group of northern Texas. Abstract. Journal of Vertebrate Paleontology **23**(3 Suppl.):58-59.

Hazel, J. E. 1969. *Cythereis eaglefordensis* Alexander, 1929; a guide fossil for deposits of latest Cenomanian age in the Western Interior and Gulf Coast regions of the United States. Pages D155-D158 *in* U.S. Geological Survey, Reston, VA. Professional Paper 650.

Hill, M. E. III. 1975. Selective dissolution of mid-Cretaceous (Cenomanian) calcareous nannofossils. Micropaleontology **21**(2):227-235.

Hook, S. C. and Cobban, W. A. 1983. Mid-Cretaceous molluscan sequence at Gold Hill, Jeff Davis County, Texas, with comparison to New Mexico. Pages 48-54 *in* Contributions to Mid-Cretaceous paleontology and stratigraphy of New Mexico Part II. New Mexico Bureau of Mines and Mineral Resources, Socorro, NM. Circular 185.

Hovorka, S. D. 1996a. High-frequency cyclicity during eustatic sea-level rise; Edwards Group of the Balcones fault zone. Bulletin of the American Association of Petroleum Geologists **80**(9):1504.

Hovorka, S. D. 1996b. High-frequency cyclicity during eustatic sea-level rise; Edwards Group of the Balcones fault zone. Gulf Coast Association of Geological Societies Transactions **46**:179-184.

Huffman, M. E. 1960. Micropaleontology of lower portion of Boquillas Formation near Hot Springs, Big Bend National Park, Brewster County, Texas. Master's Thesis. Texas Tech University, Lubbock, TX.

Humphreys, C. H. 1984a. Stratigraphy and petrography of the Lower Cretaceous (Albian) Salmon Peak Formation of the Maverick Basin, South Texas. Master's Thesis. University of Texas, Arlington, TX.

Humphreys, C. H. 1984b. Stratigraphy of the Lower Cretaceous (Albian) Salmon Peak Formation of the Maverick Basin, South Texas. Pages 34-59 *in* C. I. Smith, editor. Stratigraphy and structure of the Maverick Basin and Devils River Trend, Lower Cretaceous, Southwest Texas; a field guide and related papers. San Antonio Geological Society, San Antonio, TX.

Jacobs, L. L., K. Ferguson, M. J. Polcyn, and C. Rennison. 2005a. Cretaceous delta (super 13) C stratigraphy and the age of dolichosaurs and early mosasaurs. Geologie en Mijnbouw. Netherlands Journal of Geosciences **84**(3):257-268.

Jacobs, L. L., M. J. Polcyn, L. H. Taylor, and K. Ferguson. 2005b. Sea-surface temperatures and palaeoenvironments of dolichosaurs and early mosasaurs. Geologie en Mijnbouw. Netherlands Journal of Geosciences **84**(3):269-281.

Johnson, C. C. 2001. Taxonomic and morphologic changes in corals and rudists from Aptian and Albian strata, Texas and Puerto Rico. Geological Society of America Abstracts with Programs **33**(6):13.

Johnson, J. G. and A. S. Johnson. 2008. Prehistoric land use at Amistad National Recreation Area, Val Verde County, Texas: Report on the Texas Archeological Society 1999 Field School. Pages 1-55 *in* T. K. Pettula and J. E. Bruseth, editors. Collected papers from past Texas Archeological

Society Summer Field Schools. Texas Archeological Society, San Antonio, TX. Special Publication 5.

Johnson, J. H. 1944. Algal reefs in Cretaceous Austin Chalk of Terlingua District, Brewster County, Texas. Bulletin of the American Association of Petroleum Geologists 28(1):123-124.

Johnson, L., Jr. 1959. Palynology at Diablo Reservoir, Part I: Analysis of Sites 41VV189 and 41VV191. Report submitted to the National Park Service by the Division of Research in Anthropology. University of Texas, Austin, TX.

Johnson, L., Jr. 1963. Pollen analysis of two archeological sites at Amistad Reservoir, Texas. The Texas Journal of Science 15(2):225-230.

Jones, J. I. 1960. The significance of variability in *Praeglobotruncana gautierensis* (Broenniman), 1952, from the Cretaceous Eagle Ford group of Texas. Contributions to the Cushman Foundation for Foraminiferal Research 11:89-103.

Jones, J. O. 1993. Amistad National Recreation Area: Stratigraphy and Paleontology. Page 3 in National Park Service Paleontological Research Abstract Volume. Technical Report NPS/NRPEFO/NRTR-93/11.

Jurgens, C. J. 2005. Zooarcheology and bone technology from Arenosa Shelter (41VV99), Lower Pecos region, Texas. PhD Dissertation. University of Texas, Austin, TX.

Kaluarachchi, W., J. H. Labadie, J. Russ. 1995. A close look at the rock art of Amistad National Recreation Area, Texas. Park Science 15(4):1, 16-17.

Kariminia, S. M. 2004. Extraction of calcified Radiolaria and other calcified microfossils from micritic limestone utilizing acetic acid. Micropaleontology 50(3):301-306.

Kauffman, E. G. 1965. Middle and Late Turonian oysters of the *Lopha lugubris* group. Smithsonian Miscellaneous Collections 148(6):1-92.

KellerLynn, K. 2008. Geologic resources inventory scoping summary, Amistad National Recreation Area, Texas. National Park Service, Geologic Resources Division, Lakewood, CO. Online: http://nature.nps.gov/geology/inventory/publications/s_su mmaries/AMIS_GRE_scoping_summary_2008-0922.pdf. Accessed June 2009.

Kennedy, W. J. 1988. Late Cenomanian and Turonian ammonite faunas from North-east and central Texas. Palaeontological Association, London. Special Papers in Palaeontology 39.

Kennedy, W. J. and W. A. Cobban. 1993. Lower Cenomanian *Forbesiceras brundrettei* zone ammonite fauna in Texas, U.S.A. Neues Jahrbuch fuer Geologie und Palaeontologie. Abhandlungen 188(3):327-344.

Kennedy, W. J., W. A. Cobban, J. M. Hancock, and A. S. Gale. 2005. Upper Albian and Lower Cenomanian ammonites from the Main Street Limestone, Grayson Marl and Del Rio Clay in northeast Texas. Cretaceous Research 26(3):349-428.

Kerans, C. 2002. Styles of rudist buildup development along the northern margin of the Maverick Basin, Pecos River canyon, Southwest Texas. Gulf Coast Association of Geological Societies Transactions 52:501-506.

Kerans, C., W. Fitchen, L. Zahm, and K. Kempter. 1995. High-frequency sequence framework of Cretaceous (Albian) carbonate ramp reservoir analog outcrops, Pecos River Canyon, northwestern Gulf of Mexico Basin. University of Texas Bureau of Economic Geology, Austin, TX. Field Trip Guidebook.

Kerans, C. and R. G. Loucks. 2002. Stratigraphic setting and controls on occurrence of high-energy carbonate beach deposits; Lower Cretaceous of the Gulf of Mexico. Gulf Coast Association of Geological Societies Transactions 52:517-526.

Kerans, C. and L. C. Zahm. 1998. Facies partitioning at the high-frequency sequence scale in carbonate ramp systems. American Association of Petroleum Geologists Annual Meeting Abstracts 1998:55.

Kummel, B. 1948. Environmental significance of dwarfed cephalopods. Journal of Sedimentary Petrology 18:61-64.

Labadie, J. H. 2004. Cultural Resources Management at Amistad National Recreation Area, Del Rio, Texas. Pages 70-82 in Faces and Places of the Chihuahuan Desert 2003-2004 Speaker Series Proceedings. El Paso, TX.

Labadie, J., P. Rogers, V. Salazar-Halfmoon, M. S. Grant, and H. Young. 2005. Collection Management Plan. National Park Service, Amistad National Recreation Area.

Lock, B. E., F. S. Bases, and R. A. Glaser. 2007. The Cenomanian sequence stratigraphy of central to West Texas. Transactions of the Gulf Coast Association of Geological Societies 57:465-479.

Lock, B. E. and S. -J. Choh. 1997. Tepee structures in Boquillas Formation near Del Rio, Texas, resulting from sub-Recent calichification. Bulletin of the American Association of Petroleum Geologists 81(5):868.

Lock, B. E., S. -J. Choh, and J. J. Willis. 2001. Tepees and other surficial deformation features of Cretaceous rocks in central and West Texas, resulting from late Cenozoic caliche formation Gulf Coast Association of Geological Societies Transactions 51:173-185.

Lock, B. E., and A. W. Fife. 2004. Contourites and related outer shelf/upper slope sediments, Boquillas Formation, West Texas. American Association of Petroleum Geologists Annual Meeting Expanded Abstracts 13:85-86.

Lock, B. E. and L. Peschier. 2006. Boquillas (Eagle Ford) upper slope sediments, West Texas; outcrop analogs for potential shale reservoirs. Transactions of the Gulf Coast Association of Geological Societies 56:491-508.

Loeblich, A. R., Jr. and H. N. Tappan. 1946. New Washita foraminifera [Oklahoma, Texas]. Journal of Paleontology 20(3):238-258.

Loeblich, A. R., Jr. and H. N. Tappan. 1961. Cretaceous planktonic foraminifera; Part 1, Cenomanian. Micropaleontology 7(3):257-304.

Longoria, J. F. 1973. *Pseudoticinella*, a new genus of planktonic foraminifera from the Early Turonian of Texas. Revista Espanola de Micropaleontologia 5(3):417-423.

Lord, K. J. 1984. The zooarcheology of Hinds Cave (41VV456). PhD Disseration. Texas A&M University, College Station, TX.

Lozo, F. E. and C. I. Smith. 1964. Revision of Comanche Cretaceous stratigraphic nomenclature, southern Edwards Plateau, southwest Texas. Gulf Coast Association of Geological Societies Transactions 14:285-306.

Lundelius, E. L., Jr. 1984. A late Pleistocene mammalian fauna from Cueva Quebrada, Val Verde County, TX. Pages 456-481 *in* H. H. Genoways and M. R. Dawson, editors. Contributions in Quaternary Vertebrate Paleontology: A Volume in Memorial to John E. Guilday. Carnegie Museum of Natural History, Pittsburgh. Special Publication 8.

Maddocks, R. F. 1988. One hundred million years of predation on ostracods; the fossil record in Texas. Pages 637-657 *in* T. Hanai, N. Ikeya, and K. Ishizaki, editors. Evolutionary Biology of Ostracoda; Its Fundamentals and Applications. Elsevier, Netherlands.

Mancini, E. A. 1975. Grayson micromorph fauna (late Cretaceous). Geological Society of America Abstracts with Programs 7(2):212.

Mancini, E. A. 1977. Paedomorphism in upper Cretaceous oyster populations of Texas. Geological Society of America Abstracts with Programs 9(2):163.

Mancini, E. A. 1978a. Foraminiferal paleoecology of Grayson Formation (Upper Cretaceous) of North-central Texas. Bulletin of the American Association of Petroleum Geologists 62(9):1761.

Mancini, E. A. 1978b. Foraminiferal paleoecology of the Grayson Formation (Upper Cretaceous) of North-central Texas. Gulf Coast Association of Geological Societies Transactions 28:295-311.

Mancini, E. A. 1978c. Origin of the Grayson micromorph fauna (Upper Cretaceous) of North-central Texas. Journal of Paleontology 52(6):1294-1314.

Mancini, E. A. 1979. Late Albian and early Cenomanian Grayson ammonite biostratigraphy in North-central Texas. Journal of Paleontology 53(4):1013-1022.

Mancini, E. A. 1982. Early Cenomanian cephalopods from the Grayson Formation of north-central Texas. Cretaceous Research 3(3):241-259.

Marks, E. 1952. Occurrence of Santonian crinoid in western Gulf region. American Journal of Science 250(3):226-227.

Matthews, W. H. 1978. Texas fossils: an amateur collector's handbook. Guidebook 2. University of Texas Bureau of Economic Geology, Austin, TX.

Maudlin, R. A. 1985. Foraminiferal biostratigraphy, paleoecology, and correlation of the Del Rio Clay (Cenomanian) from Big Bend National Park, Brewster County, and Dona Ana County, New Mexico. Master's Thesis. University of Texas, El Paso, TX.

McNulty, C. L., Jr., 1964. Foraminifers from the Eagle Ford - Austin contact, northeast Texas. Bulletin of the American Association of Petroleum Geologists 48(4):537-538.

McNulty, C. L., Jr. and B. H. Slaughter. 1964. A protostegid turtle ramus from the Upper Cretaceous of Texas. Copeia 2:454.

McNulty, C. L., Jr. 1970. A new cylindracanthid rostrum from the Eagle Ford Shale (Turonian), Dallas County, Texas. The Texas Journal of Science 21(3):337.

Meyer, J. P. 1988. Shark's Tooth Hill. Lapidary Journal 42(4):65-66.

Miller, B. C. 1984. Physical stratigraphy and facies analysis, Lower Cretaceous, Maverick Basin and Devils River Trend, Uvalde and Real counties, Texas. Bulletin of the American Association of Petroleum Geologists 68(4):508.

Moreman, W. L. 1927. Fossil zones of the Eagle Ford of north Texas. Journal of Paleontology 1(1):89-101.

Moreman, W. L. 1942. Paleontology of the Eagle Ford of north and central Texas. Journal of Paleontology 16(2):192-220.

National Park Service. 1974. Natural Resources Management Plan for Amistad Recreation Area. National Park Service Southwest Region, Santa Fe, NM. Division of Natural Sciences.

Nebrigic, D. 2006. Cenomanian-Turonian bioevents-implications for sequence stratigraphic analysis. American Association of Petroleum Geologists Annual Meeting Abstracts 15:77.

Offeman, I. D., R. H. Ganshirt, P. C. Lewis, M. Martin, Jr., S. W. Arnette, R. E. Akers, T. J. Akers, and R. M. Landry. 1982. Texas Cretaceous Bivalves and Localities. Houston Gem and Mineral Society, Houston, TX. Texas Paleontology Series Publication 2.

Pampe, W. R. 1979. A dwarfed fauna from the Grayson Formation near Lake Waco, Texas. Earth Science Bulletin 12(3):18-32.

Perkins, B. F. 1951. *Hindeastraea discoidea* White from the Eagle Ford shale, Dallas County, Texas. Fondren Science Series 2.

Peschier, Lauren S. 2006. The depositional environment of Boquillas Formation, West Texas. Master's Thesis. University of Louisiana, Lafayette, LA.

Pessagno, E. A., Jr. 1969. Cenomanian-Turonian (Eaglefordian) stratigraphy in the western Gulf Coastal Plain area. Proceedings of the International Conference on Planktonic Microfossils 2:509-525.

Polcyn, M. J. and G. L. Bell, Jr. 1996. A complete skull of a new mosasauroid from the Arcadia Park Member of the Eagle Ford Shale (late middle Turonian) near Dallas, Texas. Abstract. Journal of Vertebrate Paleontology 16(3 Suppl.):58.

Powell, J. D. 1963. Turonian (Cretaceous) ammonites from northeastern Chihuahua, Mexico. Journal of Paleontology 37(6):1217-1232.

Powell, J. D. 1967. Mammitine ammonites in Trans-Pecos Texas. The Texas Journal of Science 19(3):311-322.

Price, L. L. 1931. Fishing in the Austin Chalk [*Xiphactinus audax*]. Compass of Sigma Gamma Epsilon 12(1):11-12.

Raun, G. G. and L. J. Eck. 1967. Vertebrate remains from four archeological sites in the Amistad Reservoir area, Val Verde County, Texas. The Texas Journal of Science 19(2):138-150.

Reaser, D. F. and W. C. Dawson. 1995a. Geologic study of Upper Cretaceous (Cenomanian) Buda Limestone in Northeast Texas with analysis of some regional implications. Gulf Coast Association of Geological Societies Transactions 45:495-502.

Reaser, D. F. and W. C. Dawson. 1995b. Geologic study of Upper Cretaceous (Cenomanian) Buda Limestone outliers in Northeast Texas. Bulletin of the American Association of Petroleum Geologists 79(10):1566.

Reaser, D. F. and W. C. Robinson. 2003. Cretaceous Buda Limestone in West Texas and northern Mexico. Pages 337-373 *in* R. W. Scott, editor. Cretaceous stratigraphy and paleoecology, Texas and Mexico; Perkins memorial volume. Society of Economic Paleontologists and Mineralogists, Gulf Coast Section, Houston, TX. GCSSEPM Foundation Special Publications in Geology 1.

Reinhard, K. J., S. A. de Miranda Chaves, and A. Iniguez. 2003. Plant DNA in prehistoric coprolites; evidence of diet or contamination. American Association of Stratigraphic Palynologists Annual Abstracts 27:244.

Richardson, E. S., Jr. 1955. A new variety of Cretaceous decapod from Texas. Fieldana Zoology 37:445-448.

Roemer, F. 1849. Texas, Mit besonderer Rücksicht auf deutsche Auswanderung und die physischen Verhältnisse des Landes nach eigener Beobachtung geschildert. Adoph Marcus, Bonn, Germany.

Santucci, V. L., J. Kenworthy, and R. Kerbo. 2001. An inventory of paleontological resources associated with National Park Service caves. National Park Service, Geologic Resources Division, Lakewood, CO. Technical Report NPS/NRGRD/GRDTR-01/02.

Santucci, V. L., J. P. Kenworthy, and A. L. Mims. In Press. Monitoring in situ paleontological resources. *in* R. Young and L. Norby, editors. Geological Monitoring. Geological Society of America, Boulder, Colorado. Special Paper.

Santucci, V. L., J. P. Kenworthy, and C. C. Visaggi. 2007. Paleontological Resource Inventory and Monitoring. Chihuahuan Desert Network. National Park Service TIC# D-500.

Schell, W. W. 1952. Foraminifera of the Eagle Ford shale in the type area, Dallas and Tarrant counties. Master's Thesis. Southern Methodist University, Dallas, TX.

Schmidt Murphy, S. M. 1991. Microstructure and systematic paleontology of Cenomanian stromatoporoids of the Buda Limestone, Trans-Pecos Texas. Master's Thesis. University of Texas, Arlington, TX.

Schneider, C. L. and D. R. Ruez, Jr. 2001. A paleoenvironmental conundrum; vertebrates versus invertebrates in the Late Cretaceous Eagle Ford Formation. Abstract. Journal of Vertebrate Paleontology 21(3):98.

Scotese, C. R., 2001. Atlas of Earth History, Volume 1, Paleogeography. PALEOMAP Project, Arlington, Texas.

Scott, G. 1924. Some gerontic ammonites of the Duck Creek Formation. Texas Christian University Quarterly 1:5-31.

Scott, G. 1940a. Paleoecological factors controlling distribution and mode of life of Cretaceous ammonoids in the Texas area. Journal of Paleontology 14:229-323.

Scott, G. 1940b. Pyrite faunas of Washita and Eagle Ford groups of Texas Cretaceous and the paleoecological significance. Geological Society of America Bulletin 51(12):1977.

Scott, R. W. 1990. Models and stratigraphy of mid-Cretaceous reef communities, Gulf of Mexico. SEPM Concepts in Sedimentology and Paleontology 2.

Scott, R. W. 2002a. Albian caprinid rudists from Texas re-evaluated. Journal of Paleontology 76(3):408-423.

Scott, R. W. 2002b. Upper Albian benthic foraminifers new in West Texas. Journal of Foraminiferal Research 32(1):43-50.

Scott, R. W., editor. 2007. Cretaceous rudists and carbonate platforms: Environmental feedback. Society for Sedimentary Geology, Tulsa, OK. SEPM Special Publication 87.

Shuler, E. W. 1950. A new elasmosaur from the Eagle Ford shale of Texas; the elasmosaur and its environment. Fondren Science Series 1.

Smith, C. I., B. C. Miller, and P. R. Rose. 1984. Second day road log; Uvalde to Del Rio. Pages 107-114 *in* C. I. Smith, editor. Stratigraphy and structure of the Maverick Basin and Devils River Trend, Lower Cretaceous, Southwest Texas; a field guide and related papers. San Antonio Geological Society, San Antonio, TX.

Smith, C. I. and J. B. Brown. 1983. First day road log. Pages 6-26 *in* E. C. Kettenbrink Jr., editor. Structure and stratigraphy of the Val Verde Basin-Devils River Uplift, Texas. West Texas Geological Society, Midland, TX. Publication 83-77.

Smith, C. I., J. B. Brown, and F. R. Lozo. 2000. Regional stratigraphic cross sections, Comanche Cretaceous (Fredericksburg-Washita Division), Edwards and Stockton plateaus, West Texas; interpretation of sedimentary facies, depositional cycles, and tectonics. Texas Bureau of Economic Geology, Austin, TX.

Sobolik, K. D. 1988. The prehistoric diet and subsistence of the lower Pecos region, as reflected in coprolites from Baker Cave, Val Verde County, Texas. Master's Thesis. Texas A&M University, College Station, TX.

Sobolik, K. D. 1989. The identification and quantification of pollen recovered from coprolites. American Association of Stratigraphic Palynologists Abstracts **13**:286.

Spearing, D. 1991. Roadside Geology of Texas. Mountain Press Publishing Company, Missoula, MT.

Springer, V. G. 1957. A new genus and species of elopid fish (*Laminospondylus transverses*) from the Upper Cretaceous of Texas. Copeia **2**:135-140.

Steinman, D. M. P. 1974. Cretaceous benthonic foraminifera of the Eagle Ford Formation [Texas]. Master's Thesis. University of Idaho, Moscow, ID.

Stephenson, L. W. 1929. Two new mollusks of the genera *Ostrea* and *Exogyra* from the Austin Chalk, Texas. Proceedings of the United States National Museum **76**:1-6.

Stephenson, L. W. 1944. Fossils from limestone of Buda age in Denton County, Texas. Bulletin of the American Association of Petroleum Geologists **28**(10):1538-1541.

Stephenson, L. W. 1955. Basal Eagle Ford fauna (Cenomanian) in Johnson and Tarrant counties, Texas. Pages 53-67 *in* U.S. Geological Survey, Reston, VA. Professional Paper 274.

Stewart, J. D. and V. Friedman. 2001. Oldest North American record of Saurodontidae (Teleostei, Ichthyodectiformes). Abstract. Journal of Vertebrate Paleontology **21**(3 Suppl.):104.

Stock, J. A. 1983. The prehistoric diet of Hinds Cave (41VV456), Val Verde County, Texas: the coprolite evidence. Master's Thesis. Texas A&M University, College Station, TX.

Stone, J. F. 1967. Quantitative palynology of a Cretaceous Eagle Ford exposure. Compass of Sigma Gamma Epsilon **45**(1):17-25.

Story, D. A. and V. M. Bryant, Jr. 1966. A preliminary study of the paleoecology of the Amistad Reservoir area. Final report of research sent to the National Science Foundation.

Superintendent's Compendium. 2006. Amistad National Recreation Area. National Park Service.

Tappan, H. N. 1939. Foraminifera from the Grayson Formation of northern Texas. Proceedings of the Oklahoma Academy of Science **19**(93):113.

Tappan, H. N. 1940. Foraminifera from the Grayson Formation of northern Texas. Journal of Paleontology **14**(2):93-126.

Trevino, R. H., III. 1988. Facies and depositional environments on the Boquillas Formation, Upper Cretaceous of Southwest Texas. Master's Thesis. University of Texas, Arlington, TX.

Trevino, R. H. and C. I. Smith. 2002. Facies and depositional environments of the Boquillas Formation. American Association of Petroleum Geologists Annual Meeting Abstracts. **2002**:177-178.

Udden, J. A. 1908. Fossil tracks in the Del Rio clay. Transactions and Proceedings of the Texas Academy of Science **10**:51-52.

Vega, F. J., T. Nyborg, A. Rojas-Briceno, P. Patarroyo, J. Luque, H. Porras-Muzquiz, and W. Stinnesbeck. 2007. Upper Cretaceous Crustacea from Mexico and Colombia; similar faunas and environments during Turonian times. Revista Mexicana de Ciencias Geologicas **24**(3):403-422.

Visaggi, C.C. 2006. Investigation of paleontological resources at Amistad National Recreation Area. Geological Society of America Abstracts with Programs **38**(7):35.

Webster, R. E. and G. P. Bolden. 1983. Second day road log. Pages 27-47 *in* E. C. Kettenbrink Jr., editor. Structure and stratigraphy of the Val Verde Basin-Devils River Uplift, Texas. West Texas Geological Society, Midland, TX. Publication 83-77.

Webster, R. E. and G. P. Bolden. 1984. Third day road log; Del Rio to San Antonio. Pages 115-128 *in* C. I. Smith, editor. Stratigraphy and structure of the Maverick Basin and Devils River Trend, Lower Cretaceous, Southwest Texas; a field guide and related papers. San Antonio Geological Society, San Antonio, TX.

Welles, S. P. 1949. A new elasmosaur from the Eagle Ford shale of Texas; systematic description. Fondren Science Series 1.

Welles, S. P. and B. H. Slaughter. 1963. The first record of the plesiosaurian genus *Polyptychodon* (Pliosauridae) from the New World. Journal of Paleontology 37(1):131-133.

Wells, J. W. 1934. A new species of calcisponge from the Buda limestone of central Texas. Journal of Paleontology 8(2):167-170.

Wells, J. W. 1944. A new coral from the Buda limestone (Cenomanian) of Texas. Journal of Paleontology 18(1):100.

Williams-Dean, G. J. 1978. Ethnobotany and cultural ecology of prehistoric man in southwest Texas. PhD Dissertation. Texas A&M University, College Station, TX.

Willimon, E. L. 1973. An enchodontid skull from the Austin Chalk (upper Cretaceous) of Dallas, Texas. The Southwestern Naturalist 18(2):201-210.

Willis, M. J. 1997. Paleoecology and distribution of echinoderms in the Grayson Formation, McLennan County, Texas. Bachelor's Thesis. Baylor University, Waco, TX.

Young, K. 1961. Edwards Plateau ammonite localities. Pages A3-A5 in Southern Edwards Plateau. Gulf Coast Association of Geological Societies, San Antonio, TX. Annual Meeting Guidebook.

Young, K. 1962. Ammonites of the Buda Limestone (Cretaceous), Texas and Mexico. The Texas Journal of Science 14(4):420-421.

Young, K. 1963. Upper Cretaceous ammonites from the Gulf Coast of the United States. University of Texas Bureau of Economic Geology, Austin, TX. Publication 6304.

Young, K. 1979. Edwards Plateau ammonites. Pages 59-75 *in* P. R. Rose, editor. Stratigraphy of the Edwards Group and equivalents, eastern Edwards Plateau, Texas. South Texas Geological Society, San Antonio, TX. Edition 2.

Young, K. 1984. *Hysteroceras* Hyatt [Cretaceous (Albian) ammonoid] in Texas and the Angola connection. The Texas Journal of Science 36:185-195.

Young, K. P. 1958. Cenomanian (Cretaceous) ammonites from Trans-Pecos Texas. Journal of Paleontology 32(2):286-294.

Young, K. P. and E. Marks. 1952. Zonation of Upper Cretaceous Austin Chalk and Burditt Marl, Williamson County, Texas. Bulletin of the American Association of Petroleum Geologists 36(3):477-488.

Zahm, L. C. 1997. Depositional model and sequence stratigraphic framework for upper Albian/lower Cenomanian carbonate ramp, western Comanche Shelf, Texas. Master's Thesis. University of Texas, Austin, TX.

Zahm, L. C., C. Kerans, and J. L. Wilson. 1995a. Cyclostratigraphic and ichnofacies analysis of the upper Albian Salmon Peak Formation, Maverick Basin, Texas. Bulletin of the American Association of Petroleum Geologists 79(10):1569.

Zahm, L. C., C. Kerans, and J. L. Wilson. 1995b. Cyclostratigraphic and ichnofacies analysis of the upper Albian Salmon Peak Formation, Maverick Basin, Texas. Gulf Coast Association of Geological Societies Transactions 45:595-604.

Additional References

Adams, C. S. 1988. The nature, distribution, and significance of very hard limestones in the Comanchean Series, central Texas. Bachelor's Thesis. Baylor University, Waco, TX.

Alshuaibi, A. A. 2006. Coniacian to lowermost Campanian stratigraphy of the Austin Chalk, northeast Texas. PhD Dissertation. University of Texas at Dallas, Richardson, TX.

Armstrong, A. W. 1995. The use of stable isotope ratios to investigate the relative importance of Amistad Reservoir to recharge of the McKnight and associated limestones, southwestern Val Verde County, Texas. Master's Thesis. University of Texas, San Antonio, TX.

Barrett, M. L. and J. P. Goodson, Jr. 2006. High-resolution foraminiferal biostratigraphy of Cenomanian and Turonian sandstones, Tyler County, Texas. Transactions of the Gulf Coast Association of Geological Societies 56:27-37.

Bement, L. C. 1986. Mammalian faunal and cultural remains in the late Pleistocene deposits of Bonfire Shelter, 41VV218, southwest Texas. Master's Thesis. University of Texas, Austin, TX.

Bender, R. A. and T. W. McKern. 1968. Analysis of human skeletal remains from Coontail Spin. Bulletin of the Texas Archeological Society 38:66-75.

Blair, W. F. 1950. The biotic provinces of Texas. The Texas Journal of Science 2(1):93-117.

Blome, C. D., J. R. Faith, E. W. Collins, D. E. Pedraza, and K. E. Murray. 2005. Geologic map compilation of the upper Seco Creek area, Medina and Uvalde Counties, south-central Texas. U.S. Geological Survey, Reston, VA. Open-File Report 2004-1430.

Blome, C. D., J. R. Faith, D. E. Pedraza, G. B. Ozuna, J. C. Cole, A. K. Clark, T. A. Small, and R. R. Morris. 2005. Geologic map of the Edwards Aquifer recharge zone, south-central Texas. U.S. Geological Survey, Reston, VA. Scientific Investigations Map 2873.

Bryant, V. M. Jr. 1966. Pollen analysis: its environmental and cultural implications in the Amistad Reservoir Area. Master's Thesis. University of Texas, Austin, TX.

Bryant, V. M. Jr. 1969. Late full-glacial and post-glacial pollen analysis of Texas sediments. PhD Dissertation. University of Texas, Austin, TX.

Byerly, R. M., J. R. Cooper, D. J. Meltzer, M. E. Hill Jr., and J. M. Labelle. 2005. On Bonfire Shelter (Texas) as a Paleoindian bison jump: An assessment using GIS and zooarchaeology. American Antiquity 70(4):595-629.

Chadderdon, M. F. 1984. Baker Cave, Val Verde County, Texas: The 1976 Excavations. Center for Archeological Research, University of Texas, San Antonio, TX. Special Report 13.

Chenault, D. A. and L. L. Lambert. 2005. Sequence correlation of the mid-Comanche Series, South Texas region. Bulletin of the South Texas Geological Society 46(3):13-30.

Cliff, M. B. and M. A. Nash. 2003. Archeological data recovery investigations of four Burned Rock Midden sites (41VV1892, 41VV1893, 41VV1895, and 41VV1897) Val Verde County, Texas. Texas Department of Transportation, Austin, TX. Archeological Studies Program Report 51.

Collins, E. W. 1972a. Rough Canyon SE Quadrangle, University of Texas Bureau of Economic Geology, Austin TX. STATEMAP Project Geologic Maps. 1:24,000 scale map.

Collins, E. W. 1972b. Del Rio NE Quadrangle, University of Texas Bureau of Economic Geology, Austin, TX. STATEMAP Project Geologic Maps. 1:24,000 scale map.

Collins, E. W. 1972c. Del Rio NW Quadrangle. University of Texas Bureau of Economic Geology, Austin, TX. STATEMAP Project Geologic Maps. 1:24,000 scale map.

Collins, M. B. 1969. Test Excavations at Amistad International Reservoir, Fall 1967. Papers of the Texas Archeological Salvage Project No. 16. University of Texas, Austin, TX.

Dawson, W. C. 1986. Austin Chalk and Buda Limestone (Cretaceous) petroleum reservoirs in Caldwell County, Texas; a case history. Energy Exploration and Exploitation 4(5):377-389.

Denison, R. E., N. R. Miller, R. W. Scott, and D. F. Reaser. 2003. Strontium isotope stratigraphy of the Comanchean Series in North Texas and southern Oklahoma. Geological Society of America Bulletin 115(6):669-682.

Dibble, D. S. 1967. Excavations at Arenosa Shelter, 1965-1966. Mimeograph report submitted to the National Park Service by the Texas Archeological Salvage Project. University of Texas, Austin, TX.

Dibble, D. S. and R. K. Alexander. 1972. The Archeology of Texas Caves. Pages 133-148 in Natural History of Texas Caves. Dallas, TX.

Dibble, D. S. and E. R. Prewitt. 1967. Survey and Test Excavations at Amistad Reservoir, 1964-1965. Texas Archeological Salvage Project Survey Reports No. 3. University of Texas, Austin, TX.

Douglas, C. L. 1969. Catfish spines from archeological sites in Texas. Bulletin of the Texas Archeological Society **40**:263-265.

Edwards, S. K. 1990. Investigations of Late Archaic Coprolites: pollen and macrofossil remains from Hinds Cave, Val Verde County, Texas. Master's Thesis. Texas A&M University. College Station, TX.

Epstein, J. F. 1960. Centipede and Damp Caves: Excavations in Val Verde County, Texas, 1958. Report submitted to the National Park Service by the Texas Archeological Salvage Project, University of Texas, Austin, TX.

Erdogan, S. Z. 1969. Cenomanian Buda limestone (Comanche Cretaceous), of west and Trans-Pecos, Texas. Master's Thesis. Louisiana State University, Baton Rouge, LA.

Ewing, T. 1996. Summary of the geology of the San Antonio area. Pages 7-54 *in* Ewing, T., editor. Rocks, Landscapes and Man; Urban Geology of the San Antonio Area. South Texas Geological Society, San Antonio, TX. Guidebook.

Ewing, T. E. 2005. Phanerozoic development of the Llano Uplift. Bulletin of the South Texas Geological Society **45**(9):15-25.

Faith, J. R. and C. D. Blome. 2006. Fault geometry and lithofacies controls on the hydrostratigraphic framework of the Edwards Aquifer, South-Central Texas. Geological Society of America Abstracts with Programs **38**(7):26.

Freeman, V. L., 1965. Geologic map of the Baker's Crossing Quadrangle, Val Verde County, Texas. U.S. Geological Survey, Reston, VA. Miscellaneous Geologic Investigations Map I-434. 1:62,500 scale map.

Freeman, V. L. 1968. Geology of the Comstock-Indian Wells area, Val Verde, Terrell, and Brewster Counties, Texas. U.S. Geological Survey, Reston, VA. Professional Paper 594-K. 1:250,000 scale map.

Goodfriend, G. A. and G. L. Ellis. 2000. Stable carbon isotope record of middle to late Holocene climate changes from land snail shells at Hinds Cave, Texas. Quaternary International **67**:47-60.

Gustavson, T. C. and M. B. Collins. 1998. Geoarcheological Investigations of Rio Grande Terrace and Flood Plain Alluvium from Amistad Dam to the Gulf of Mexico. Texas Archeological Research Laboratory, University of Texas, Austin, TX. Technical Series 49.

Hancock, J. M. and I. Walaszczyk. 2004. Mid-Turonian to Coniacian changes of sea level around Dallas, Texas. Cretaceous Research **25**(4):459-471.

Hoadley, C.R. 1986. Paleoecology of encrusting epifauna of echinoids and oysters of the Mid-Cretaceous. Master's Thesis. Baylor University, Waco, TX.

Hixon, S. B. 1959. Facies and petrography of the Cretaceous Buda Limestone of Texas and northern Mexico. Master's Thesis. University of Texas, Austin, TX.

Hudson, R. M. 1986. Progradation of a secondary shelf margin, upper Salmon Peak Formation, Maverick Basin, Southwest Texas. Master's Thesis. University of Texas, Arlington, TX.

Heubner, J. 1991. Cactus for dinner, again! An isotopic analysis of Late Archaic diet in the Lower Pecos Region of Texas. Pages 175-190 *in* S. Turpin, editor. Papers on Lower Pecos Prehistory. Texas Archeological Research Laboratory, University of Texas, Austin, TX. Studies in Archeology 8.

Hill, R. T. 1891 Notes on the Geology of the Southwest. American Geologist **7**: 254-255, 336-370.

Jacobs, L. L. and D. A. Winkler. 1998. Mammals, archosaurs, and the Early to Late Cretaceous transition in north-central Texas. Pages 253-280 *in* Y. Tomida, L. J. Flynn, and L. L. Jacobs, editors. Advances in Vertebrate Paleontology and Geochronology. Natural Science Museum Monographs 14.

Jaroska, R. S. 1955. Population analysis of the Del Rio Formation in central and Southwest Texas. Master's Thesis. University of Texas, Austin, TX.

Johnson, L., Jr. 1964. The Devil's Mouth Site: A Stratified Campsite at Amistad Reservoir, Val Verde County, Texas. Department of Anthropology, University of Texas, Austin, TX. Archeology Series 6.

Johnston, C. 1962. Stop no. 1; The Del Rio clay. Pages 22-23 *in* Southwestern McLennan County and eastern Coryell County. Baylor Geological Society, Waco, TX. Popular Geology of Central Texas 7.

Lehman, T. M. 1985. Stratigraphy, sedimentology, and paleontology of Upper Cretaceous (Campanian-Maastrichtian) sedimentary rocks in Trans-Pecos Texas. PhD Dissertation. University of Texas, Austin, TX.

Lehman, T. M. 1986. Late Cretaceous sedimentation in Trans-Pecos Texas. Bulletin of the West Texas Geological Society **25**(7):4-9.

Lehmann, C. T. 1997. Sequence stratigraphy and platform evolution of Lower Cretaceous (Barremian-Albian) carbonates of northeastern Mexico. PhD Dissertation. University of California at Riverside, Riverside, CA.

Lock, B. E. 1996. Stratigraphic significance of diastemic surfaces; examples from the Cretaceous of west-central Texas. Geological Society of America Abstracts with Programs **28**(7):124.

Mancini, E. A. 1979. Models for interpretation of micromorph faunas in Washita Group. Bulletin of the American Association of Petroleum Geologists **63**(3):490.

Mancini, E. A. 1982. Foraminiferal population changes in a shallow epicontinental marine carbonate-claystone sequence; Main Street-Grayson interval (Cretaceous), North-central Texas. Proceedings of the North American Paleontological Convention 3:353-358.

Mancini, E. A. 2007. Sequence stratigraphy of Jurassic and Cretaceous strata, Gulf Coastal Plain. Geological Society of America Abstracts with Programs **39**(6):150.

Mancini, E. A. and R. W. Scott. 2006. Sequence stratigraphy of Comanchean Cretaceous outcrop strata of Northeast and South-Central Texas; implications for enhanced petroleum exploration. Transactions of the Gulf Coast Association of Geological Societies **56**:539-550.

McCormick, C. L., C. L. Smith, and C. D. Henry. 1996. Cretaceous stratigraphy. Pages 30-46 *in* C. D. Henry, W. R. Muehlberger, C. L. McCormick, L. L. Davis, R. J. Erdlac, Jr., M. J. Kunk, and C. I. Smith, editors. Geology of the Solitario Dome, Trans-Pecos Texas; Paleozoic, Mesozoic, and Cenozoic sedimentation, tectonism, and magmatism. University of Texas Bureau of Economic Geology, Austin, TX. Report of Investigations 240.

Mehalchick, G., T. Meyers, K. W. Kibler, and D. K. Boyd. 1999. Val Verde on the Sunny Rio Grande: Geoarcheological and Historical Investigations at San Felipe Springs, Val Verde County, Texas. Prewitt and Associates, Inc. Austin, TX. Reports of Investigations 122.

Meyer, J. P. 1991. Harvesting Cretaceous crabs. Lapidary Journal **45**(6):61-62, 64, 66, 68.

Moreman, W. L. 1925. Micrology of the Woodbine, Eagle Ford, and Austin Chalk. Pages 74-78 **in** University of Texas. Bulletin 2544.

Moreman, W. L. 1931. Stratigraphy and paleontology of the Eagle Ford Formation of north and central Texas. PhD Dissertation. University of Kansas, Lawrence, KS.

Neck, R. W. 1990. Geological substrate and human impact as influences on bivalves of Lake Lewisville, Trinity River, Texas. The Nautilus **104**(1):16-25.

North, G. R., G. Bomar, J. Griffiths, J. Norwine, and J. B. Valdes. 1995. The Changing Climate of Texas. Pages 24-49 *in* G. R. North, J. Schmandt, and J. Clarkson, editors. The Impact of Global Warming on Texas. University of Texas Press, Austin, TX.

Nunley, J. P., L. F. Duffield, and E. B. Jelks. 1965. Excavations at Amistad Reservoir, 1962 Season., Texas Archeological Salvage Project. University of Texas, Austin, TX. Miscellaneous Papers 3.

Parson, M. L. 1965. 1963 Test excavations at Fate Bell Shelter, Amistad Reservoir, Val Verde County, Texas. Texas Archeological Salvage Project, University of Texas, Austin, TX. Miscellaneous Papers 4.

Patton, J. L. 1932. The paleontology of the Austin Chalk in Travis and Williamson counties, Texas. Master's Thesis. University of Texas, Austin, TX.

Peterson, J. D. 1985. A petrologic and depositional study of the Lower Cretaceous Buda Limestone, central Texas. Bachelor's Thesis. Baylor University, Waco, TX.

Powell, J. D. 1968. Woodbine-Eagle Ford transition, Tarrant Member. Pages 27-43 *in* Stratigraphy of the Woodbine Formation. Geological Society of America South-Central Section Field Trip Guidebook.

Prewitt, E. R. 1966. A Preliminary Report on the Devil's Rockshelter Site, Val Verde County, Texas. Texas Journal of Science **18**(2):206-224.

Prewitt, E. R. and D. A. Lawson. 1972. An assessment of the archeological and paleontological resources of Lake Texoma, Texas-Oklahoma. University of Texas, Austin, TX. Texas Archeological Salvage Project Survey Reports 10.

Redfield, R. C. 1940. A study of the lower portion of the Boquillas Formation, Brewster County, Texas. Master's Thesis. University of Texas, Austin, TX.

Rodda, P. U., W. L. Fisher, W. R. Payne, and D. A. Schofield. 1966. Limestone and dolomite resources, Lower Cretaceous rocks, Texas. University of Texas Bureau of Economic Geology, Austin, TX. Report of Investigations 56.

Ross, R. E. 1965. The archeology of Eagle Cave. University of Texas, Austin, TX. Papers of the Texas Archeological Salvage Project 7.

Russ, J., R. L. Palma, and J. L. Booker. 1994. Whewellite rock crusts in the Lower Pecos region of southwest Texas. Texas Journal of Science **46**:165-172.

Russ, J., R. L. Palma, D. H. Loyd, D. W. Farwell, and H. G. M. Edwards. 1995. Analysis of the rock accretions in the Lower Pecos region of southwest Texas. Geoarchaeology **10**:43-63.

Sanders, R. B. and T. M. Lehman. 1989. Sedimentology and stable isotope geochemistry of the Ernst Member of the Boquillas Formation (Late Cretaceous) Trans-Pecos Texas. Geological Society of America Abstracts with Programs **21**(1):39.

Saunders, J. W. 1986. The Economy of Hinds Cave. PhD Dissertation. Southern Methodist University, Dallas, Texas.

Saunders, J. W. 1992. Plant and Animal Procurement Sites in the Lower Pecos Region, Texas. Journal of Field Archaeology **19**:335-349.

Schuetz, M. K. 1956. An analysis of Val Verde County Cave Materials: Part I. Bulletin of the Texas Archeological Society **27**:129-160.

Schuetz, M. K. 1961. An analysis of Val Verde County Cave Materials: Part II. Bulletin of the Texas Archeological Society **31**:167-205.

Schuetz, M. K. 1963. An analysis of Val Verde County Cave Materials: Part III. Bulletin of the Texas Archeological Society **33**:131-165.

Scott, R. W. and C. Kerans. 2004. Late Albian carbonate platform chronostratigraphy Devils River Formation cycles, west Texas. Courier Forschungsinstitut Senckenberg **247**:129-148.

Scott, R. W. and E. J. Kidson. 1977. Lower Cretaceous depositional systems, West Texas. Pages 169-181 *in* D. G. Bebout and R. G. Loucks, editors. Cretaceous carbonates of Texas and Mexico; applications to subsurface exploration. University of Texas Bureau of Economic Geology, Austin, TX. Report of Investigations 89.

Shafer, H. J. and V. M. Bryant, Jr. 1977. Archeological and Botanical Studies at Hinds Cave, Val Verde County, Texas. Annual Report to the National Science Foundation by the Department of Anthropology, Texas A&M University, College Station, TX.

Sharps, J. A. 1963. Geologic Map of the Malvado Quadrangle, Terrell and Val Verde Counties, Texas. U.S. Geological Survey, Reston, VA. Miscellaneous Geologic Investigations Map I-382. 1:62,500 scale map.

Sharps, J. A. and Freeman, V. L. 1965. Geologic map of the Mouth of Pecos and Feely Quadrangles, Val Verde County, Texas. U.S. Geological Survey, Reston, VA. Miscellaneous Geologic Investigations Map I-440. 1:62,500 scale map.

Simmons, T. L. 2007. Bioarchaeological analysis of commingled skeletal remains from Bee Cave Rockshelter (41VV546), Val Verde County, Texas. Master's Thesis. Texas State University, San Marcos, TX.

Smith, C. I. 1970. Lower Cretaceous Stratigraphy, Northern Coahuila, Mexico. University of Texas Bureau of Economic Geology, Austin, TX. Report of Investigations 65.

Smith, C. I., B. C. Miller, and P. R. Rose. 1984. First day road log; San Antonio to Uvalde. Pages 101-106 *in* C. I. Smith, editor. Stratigraphy and structure of the Maverick Basin and Devils River Trend, Lower Cretaceous, Southwest Texas; a field guide and related papers. San Antonio Geological Society, San Antonio, TX.

Stanford, J. W. 1958. The Del Rio Clay. Pages 64-68 in Guide to the mid-Cretaceous geology of central Texas. Baylor Geological Society, Waco, TX.

Stanton, G. P., W. H. Kress, A. P. Teeple, M. L. Greenslate, and A. K. Clark. 2007. Geophysical analysis of the Salmon Peak Formation near Amistad Reservoir Dam, Val Verde County, Texas, and Coahuila, Mexico, March 2006, to aid in piezometer placement. U.S. Geological Survey, Reston, VA. Scientific Investigations Report 2007-5143.

Stevens, J. B., R. A. Ashmore, L. C. Doornbos, D. A. Reed, and M. S. Stevens. 1995. Faunal change in cyclic deposits of the late Coniacian-early Santonian, Terlingua Group, Big Bend National Park, Trans-Pecos Texas. Geological Society of America Abstracts with Programs **27**(6):176.

Taylor, D. R. 1960. The Buda limestone. Pages 98-99 *in* Cretaceous stratigraphy of the Grand and Black Prairies, east-central Texas. Baylor Geological Society, Waco, TX. Field Conference Guidebook.

Taylor, W. W. and F. G. Rul. 1961. An archeological reconnaissance behind the Diablo Dam, Coahuila, Mexico. Bulletin of the Texas Archeological Society **31**:153-165.

Thomas, R. G. 1971. The geomorphic evolution of the Pecos River system. Master's Thesis. Baylor University, Waco, TX.

Thomas, S. J. 1933. The Archeological Investigation of Fate Bell Shelter, Seminole Canyon, Val Verde County, Texas. Master's Thesis. University of Texas, Austin, TX.

Van Devender, T. R. 1990. Late Quaternary Vegetation and Climate of the Chihuahuan Desert, United States and Mexico. Pages 105-133 *in* J. L. Betancourt, T. R. Van Devender, and P. S. Martin, editors. Packrat Middens: The Last 40,000 Years of Biotic Change. University of Arizona Press, Tucson, AZ.

Webster, R. E., C. F. Miller, and D. F. Reaser. 1978. Structural geology of Southwest Edwards Plateau region, Texas. Bulletin of the American Association of Petroleum Geologists **62**(3):571.

Whitney, F. L. 1911. Fauna of the Buda Limestone (Lower Cretaceous, Texas). Master's Thesis. Cornell University, Ithaca, NY.

Word, J. H. and C. L. Douglas, 1970. Excavations at Baker Cave, Val Verde County, Texas. Texas Memorial Museum, Austin, TX. Bulletin 16.

Yates, R. G. and G. A. Thompson. 1959. Geology and quicksilver deposits of the Terlingua District, Texas. U.S. Geological Survey, Reston, VA. Professional Paper 312.

Young, K. 1977. Middle Cretaceous rocks of Mexico and Texas. Pages 325-332 *in* D. G. Bebout and R. G. Loucks, editors. Cretaceous carbonates of Texas and Mexico; applications to subsurface exploration. University of Texas Bureau of Economic Geology, Austin, TX. Report of Investigations 89.

Appendix A: Fossil taxa list

This appendix lists fossil taxa recorded within AMIS. They are divided into two time periods, those that were discovered in Quaternary-aged deposits (Pleistocene-Holocene) and those that were discovered in the Cretaceous-aged deposits.

Fossil Plants [Quaternary]

Asteraceae/Compositae
 Ambrosia
 Artemisia
 Helianthus
 Iva
 Xanthium
Agavaceae
 Agave
Berberidaceae
 Berberis
Betulaceae
 Alnus
Brassicaceae/Cruciferae
Cactaceae
 Ariocarpus
 Opuntia
 Echinocereus
 Mammillaria
Chenopodiaceae
 Amaranthus
 Atriplex
 Chenopodium
Cucurbitaceae
 Cucurbita
Cupressaceae
 Juniperus
Cyperaceae
 Carex
Ebenaceae
 Diospyros
Ephedraceae
 Ephedra
Euphorbiaceae
 Jatropha
Fabaceae/Leguminosae
 Acacia
 Cassia
 Dalea
 Leucaena
 Mimosa?
 Petalostemum?
 Prosopis
 Sophora
Fagaceae
 Quercus
Fouquieriaceae
 Foquieria
Geraniaceae
 Erodium
Hamamelidaceae
 Liquidambar
Juglandaceae
 Carya

Juglans
Koeberliniaceae
 Koeberlinia
Lamiaceae/Labiatae
Linguliflorae
Liliaceae
 Allium
 Dasylirion
 Yucca
Malvaceae
 Sphaeralcea
 Tilia
Moraceae
 Maclura
 Morus
Nyctaginaceae
 Abronia
Onagraceae
 Gaura
 Jussiaea
 Oenothera
Oleaceae
 Fraxinus
Pinaceae
 Picea
 Pinus
 Pseudotsuga
Poaceae/Gramineae
 Aristida
 Cenchrus
 Panicum?
 Pappophorum
 Setaria
 Sporobolus
 Tripsacum
 Zea
Polemoniaceae
 Gilia
Polypodiaceae
Polygonaceae
 Eriogonum
Portulacaceae
 Portulaca
Ranunculaceae
 Clematis
Rhamnaceae
 Karwinskia
Ruscaceae
 Nolina
Rutaceae
 Ptelea
Salicaceae
 Populus
Sapindaceae
 Ungnadia

Fossil Plants [Quaternary], continued
    Selaginellaceae
       Selaginella
    Scrophulariaceae
       Leucophyllum
    Solanaceae
    Ulmaceae
       Celtis
       Ulmus
    Umbelliferae
    Typhaceae
       Typha
    Vitaceae
       Vitis
    Zygophyllaceae
       Larrea

Fossil Invertebrates [Quaternary and Cretaceous]

Protists
<u>Sarcomastigophora [Cretaceous]</u>
  Foraminiferida
    Globigerinidae
      Globigerina
    Miliolidae
    Textulariidae
      Cribratina texana

Animals
<u>Arthropoda [Quaternary]</u>
  Diplopoda
  Insecta
    Coeloptera
      Niptus
    Diptera
    Lepidoptera
    Orthoptera

<u>Arthropoda [Cretaceous]</u>
  Crustacea
  Ostracoda

<u>Bryozoa [Cretaceous]</u>

<u>Cnidaria [Cretaceous]</u>
  Anthozoa
  Scleractinia

<u>Echinodermata [Cretaceous]</u>
  Echinoidea
  Irregularia
  Regularia
    Salenia
  Ophiuroidea

<u>Mollusca [Quaternary]</u>
  Bivalvia
    Amblema
    Lampsilis
    Physa
    Pisidium
    Proptera

    Sphaerium
Gastropoda
  Achatinid
  Bulimulus
  Catinella
  Discus
  Durangonella
  Gastrocopta
  Gyraulus
  Hawaiia
  Helicodiscus
  Helisoma
  Laevapex
  Lamellaxis
  Planorbis
  Polygyra
  Punctum
  Pupoides
  Succinea
  Tropicorbis

<u>Mollusca [Cretaceous]</u>

Bivalvia
  Hippuritoida
  Caprinidae
  Radiolitidae
  Limoida
  Limidae
    Lima
  Pectinoida
  Pectinidae
    Neithea
  Pteroida
  Chondrodontidae
  Inoceramidae
    Inoceramus
  Gryphaeidae
    Exogyra
    Ilymatogyra
  Ostreidae
    Ostrea
  Trigonioida
  Trigoniidae
    Trigonia
  Veneroida
  Cardiidae
    Protocardia

Cephalopoda
  Nautiloidea
    Cymatoceras
  Ammonoidea
    Pervinquieria
    Plesioturrilites

Gastropoda
    Nerinea
    Tylostoma

Fossil Vertebrates [Quaternary]

Actinopterygii
  Acipenseriformes
    Scaphirhynchus
  Cypriniformes
    Carpiodes
    Catastomus
    Cycleptus
    Ictiobus
  Lepisosteiformes
    Lepisosteus
  Siluriformes
    Ictaclurus
    Pylodictus
  Perciformes
    Aplodinotus
    Micropterus
    Morone

Amphibia
  Anura
    Rana

Aves
  Anseriformes
    Aythya
    Anas
    Branta
    Chen
    Melanitta?
  Charadriiformes
  Columbiformes
    Zenaida
    Zenaidura
  Cuculiformes
    Geococcyx
  Falconiformes
    Buteo
    Falco
    Ictinia?
  Galliformes
    Colinus
    Callipepla?
    Meleagris

Mammalia
  Artiodactyla
    Antilocapridae
      Antilocapra
      Stockoceros
    Bovidae
      Bison
      Capra?
      Ovis
    Camelidae
      Camelops
    Cervidae
      Navahoceros
      Odocoileus
  Carnivora
    Canidae
      Canis
      Urocyon
      Vulpes
    Felidae
      Lynx
    Mustelidae
      Conepatus?
      Mephitis
      Mustela
      Spilogale
      Taxidea
    Procyonidae
      Bassariscus
      Procyon
    Ursidae
      Arctodus
  Lagomorpha
    Lepus
    Sylvilagus
  Perissodactyla
    Equus
  Proboscidea
    Elephas
  Rodentia
    Ammospermophilus
    Baiomys
    Castor
    Citellus
    Cratogeomys
    Erithizon
    Geomys
    Neotoma
    Ondatra
    Onychomys
    Pappogeomys
    Perognathus
    Peromyscus
    Reithrodontomys
    Sciurus
    Sigmodon
    Spermophilus
    Thomomys

Reptilia
  Chelonia
    Chelydra
    Pseudomys
    Terrapene
    Trionyx
  Sauria
    Ophisaurus
  Squamata
    Agkistrodon
    Crotalus
    Pituophis

Trace Fossils [Cretaceous]

(Annelida)
  Planolites?
  Skolithos
(Crustacea)
  Thalassinoides
(Porifera)
  Entobia

Appendix B: Map of paleontological localities

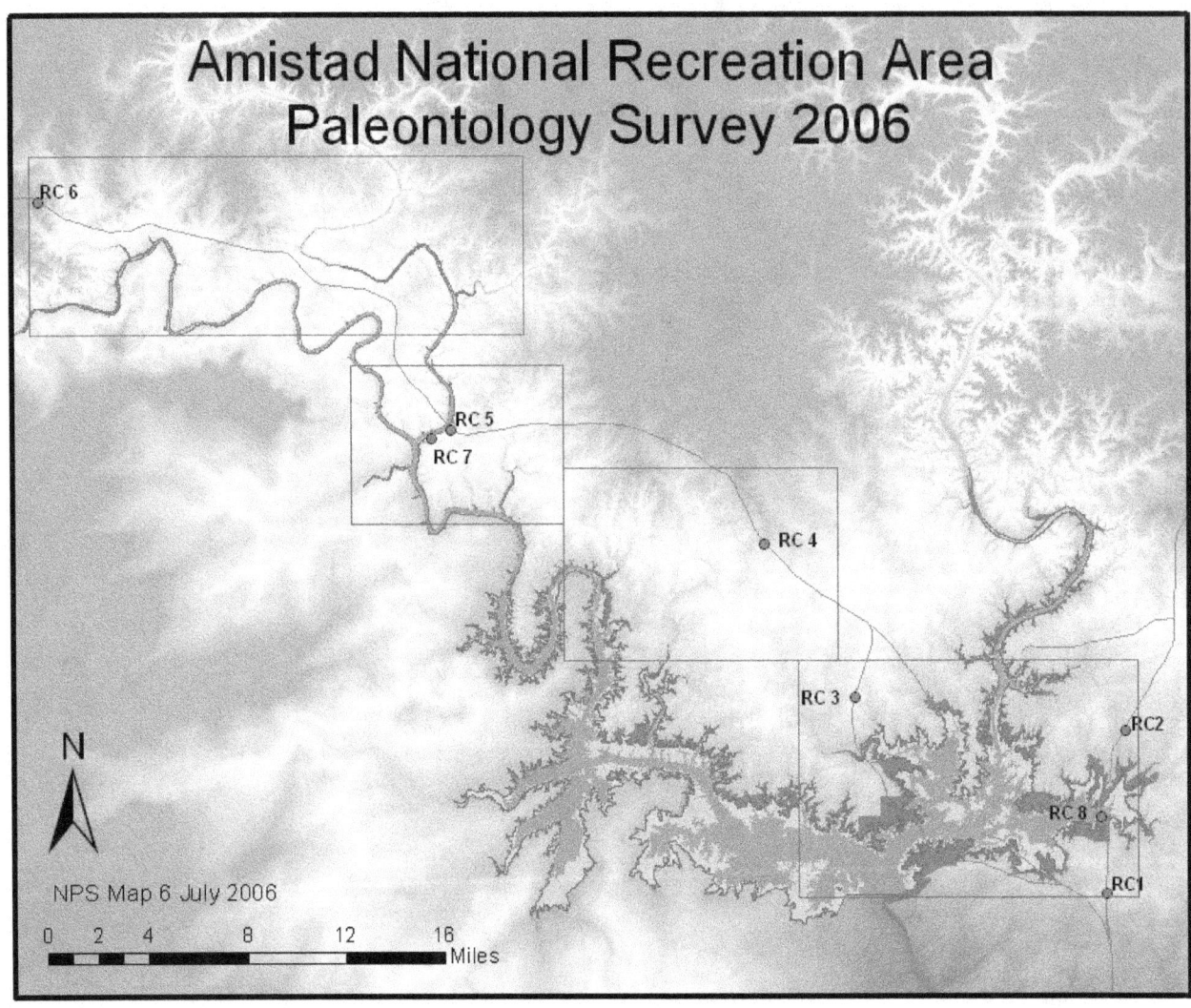

This map illustrates the locations of select fossiliferous exposures near Amistad National Recreation Area. Numbers refer to road cuts described in the *Localities* section of this report. AMIS localities are not plotted, in accordance with National Park Service policy regarding paleontological resource localities. More detailed location and geologic maps are provided in Appendix C. Map produced by AMIS GIS Staff, June 2006.

Appendix C: Maps of paleontological localities by region

This section contains regional maps with select paleontological localities indicated. The base maps are extracted from Barnes (1977). AMIS localities are not plotted on these maps. For a reference map of paleontological localities in the Amistad National Recreation Area vicinity, refer to Appendix B. Maps produced by AMIS GIS Staff, June 2006.

Diablo East, Evans and California Creek, and San Pedro Area

Map Key (Barnes, 1977):
Cenozoic

Qal	Alluvium (Recent)
Qt	Fluviatile Terrace (Pleistocene)
T-Qu	Uvalde Gravel (Pliocene or Pleistocene)

Mesozoic

Kau	Austin Chalk (Upper Cretaceous)	Kdr	Del Rio Clay (Upper Cretaceous)
Kbo	Boquillas Flags (Upper Cretaceous)	Kbd	Undivided Kbu/Kdr (Upper Cretaceous)
Kef	Eagle Ford Group (Upper Cretaceous)	Kdvr	Devils River Limestone (Lower Cretaceous)
Kbu	Buda Limestone (Upper Cretaceous)	Ksa	Salmon Peak Limestone (Lower Cretaceous)

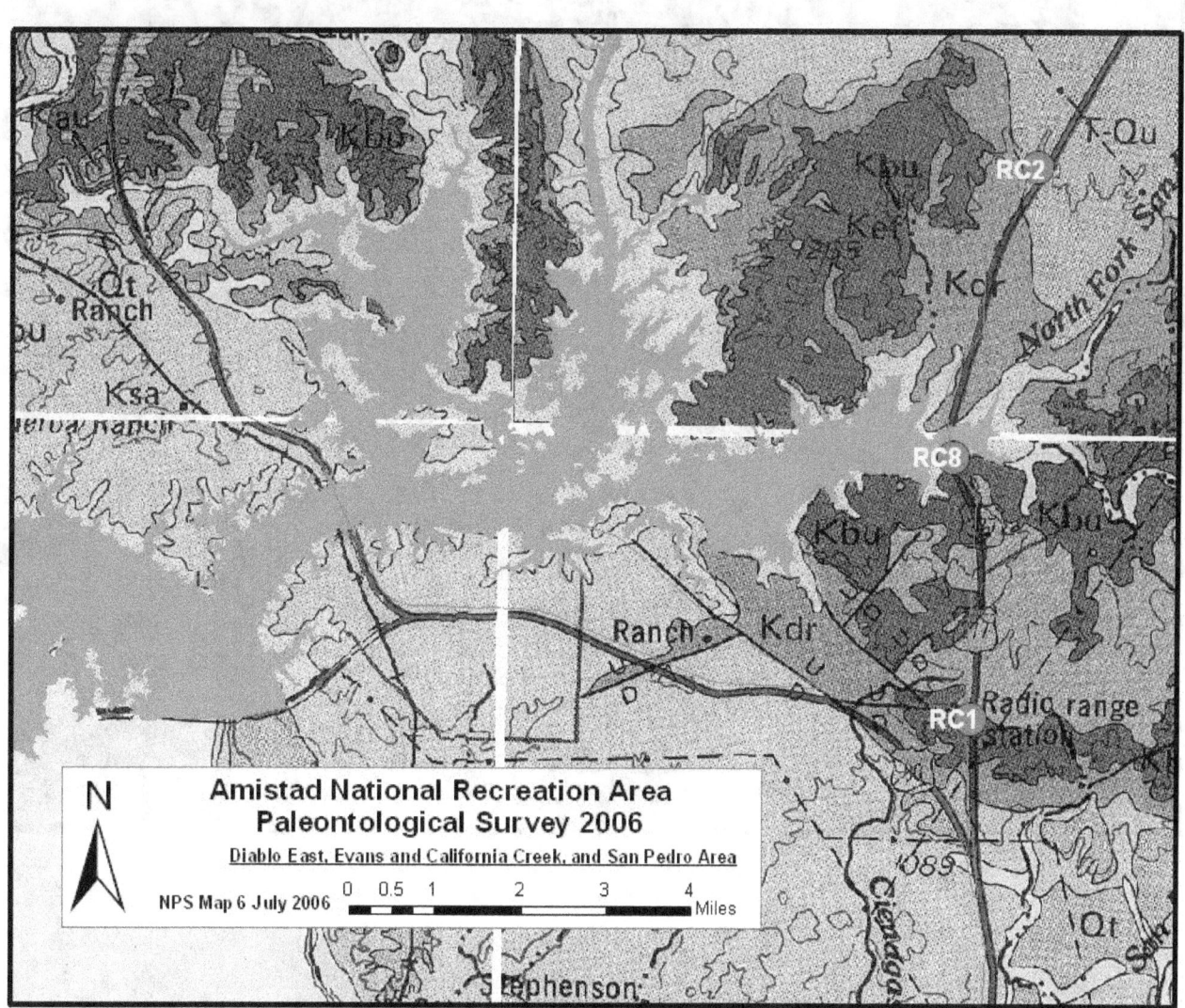

Amistad National Recreation Area
Paleontological Survey 2006
Diablo East, Evans and California Creek, and San Pedro Area
NPS Map 6 July 2006   0 0.5 1   2   3   4
Miles

51

Rio Grande and Pecos Confluence

Map Key (Barnes, 1977):
Cenozoic

Qal	Alluvium (Recent)
Qt	Fluviatile Terrace (Pleistocene)
T-Qu	Uvalde Gravel (Pliocene or Pleistocene)

Mesozoic

Kau	Austin Chalk (Upper Cretaceous)	Kdr	Del Rio Clay (Upper Cretaceous)
Kbo	Boquillas Flags (Upper Cretaceous)	Kbd	Undivided Kbu/Kdr (Upper Cretaceous)
Kef	Eagle Ford Group (Upper Cretaceous)	Kdvr	Devils River Limestone (Lower Cretaceous)
Kbu	Buda Limestone (Upper Cretaceous)	Ksa	Salmon Peak Limestone (Lower Cretaceous)

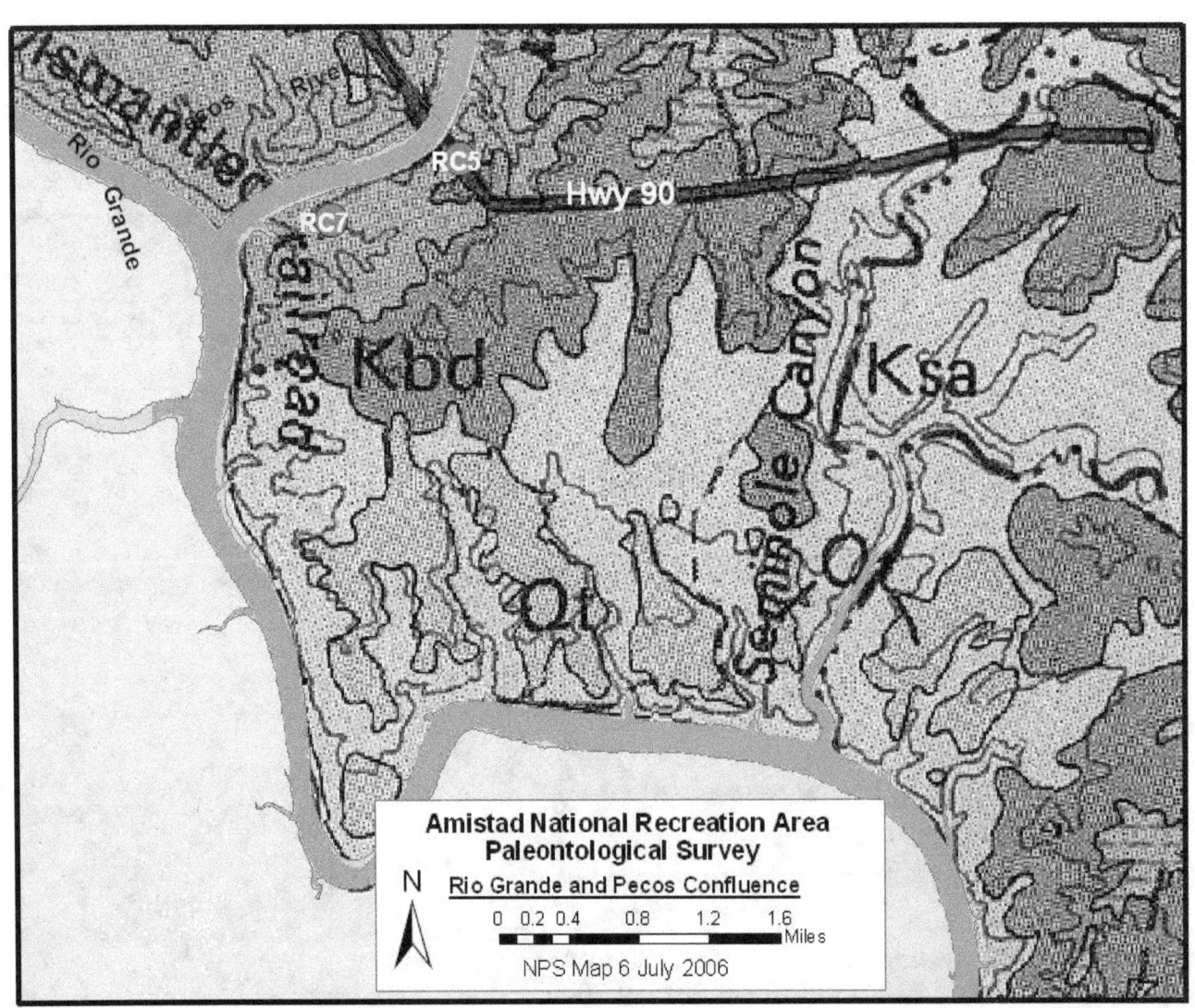

Amistad National Recreation Area
Paleontological Survey
Rio Grande and Pecos Confluence

N

0 0.2 0.4   0.8   1.2   1.6
Miles

NPS Map 6 July 2006

Highway 90 at Spur 406 and West to the Lake

Map Key (Barnes, 1977):
Cenozoic

Qal   Alluvium (Recent)
Qt   Fluviatile Terrace (Pleistocene)
T-Qu   Uvalde Gravel (Pliocene or Pleistocene)

Mesozoic

Kau   Austin Chalk (Upper Cretaceous)   Kdr   Del Rio Clay (Upper Cretaceous)
Kbo   Boquillas Flags (Upper Cretaceous)   Kbd   Undivided Kbu/Kdr (Upper Cretaceous)
Kef   Eagle Ford Group (Upper Cretaceous)   Kdvr   Devils River Limestone (Lower Cretaceous)
Kbu   Buda Limestone (Upper Cretaceous)   Ksa   Salmon Peak Limestone (Lower Cretaceous)

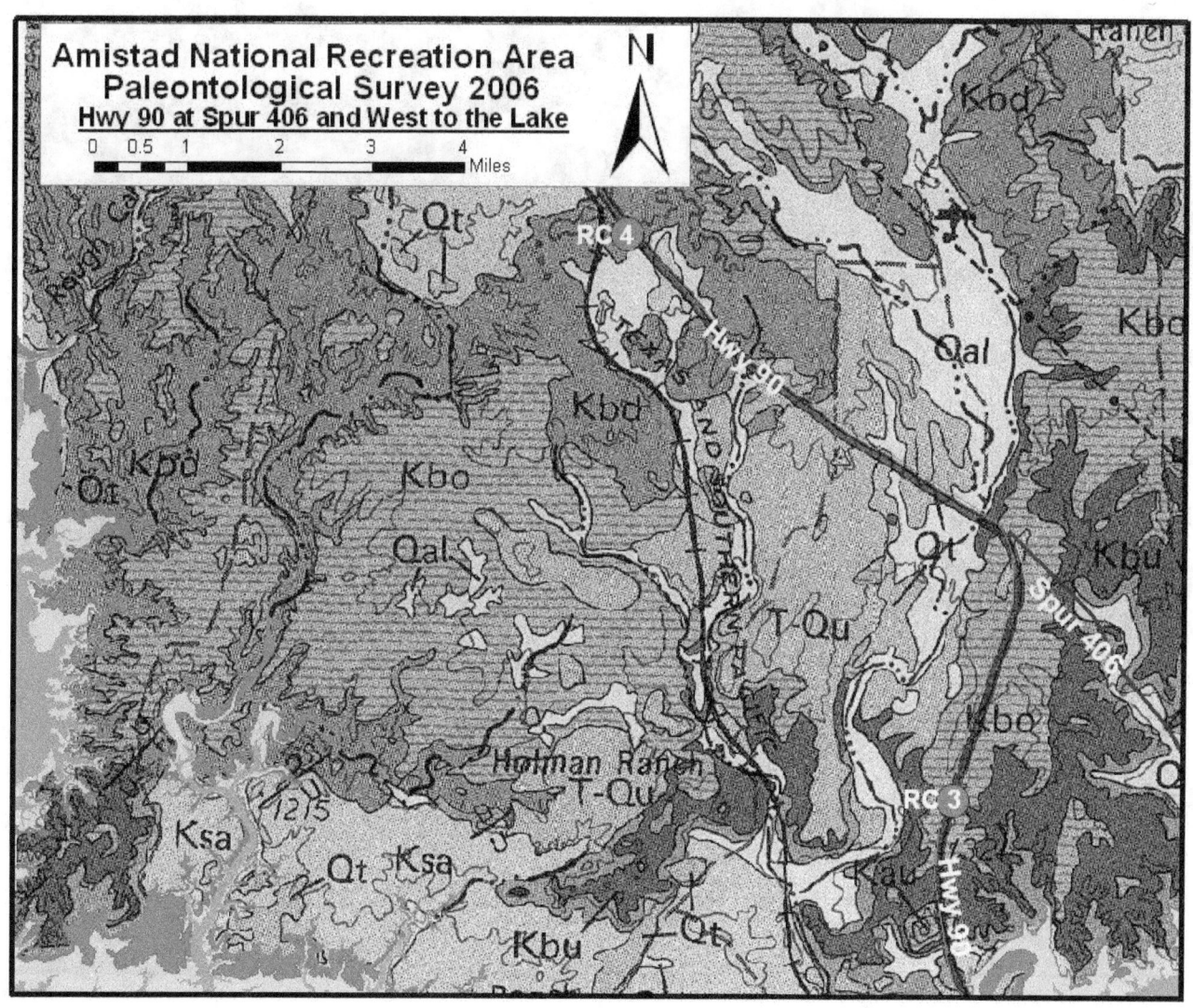

Shumla Bend and West along Rio Grande

Map Key (Barnes, 1977):
Cenozoic

Qal	Alluvium (Recent)
Qt	Fluviatile Terrace (Pleistocene)
T-Qu	Uvalde Gravel (Pliocene or Pleistocene)

Mesozoic

Kau	Austin Chalk (Upper Cretaceous)	Kdr	Del Rio Clay (Upper Cretaceous)
Kbo	Boquillas Flags (Upper Cretaceous)	Kbd	Undivided Kbu/Kdr (Upper Cretaceous)
Kef	Eagle Ford Group (Upper Cretaceous)	Kdvr	Devils River Limestone (Lower Cretaceous)
Kbu	Buda Limestone (Upper Cretaceous)	Ksa	Salmon Peak Limestone (Lower Cretaceous)

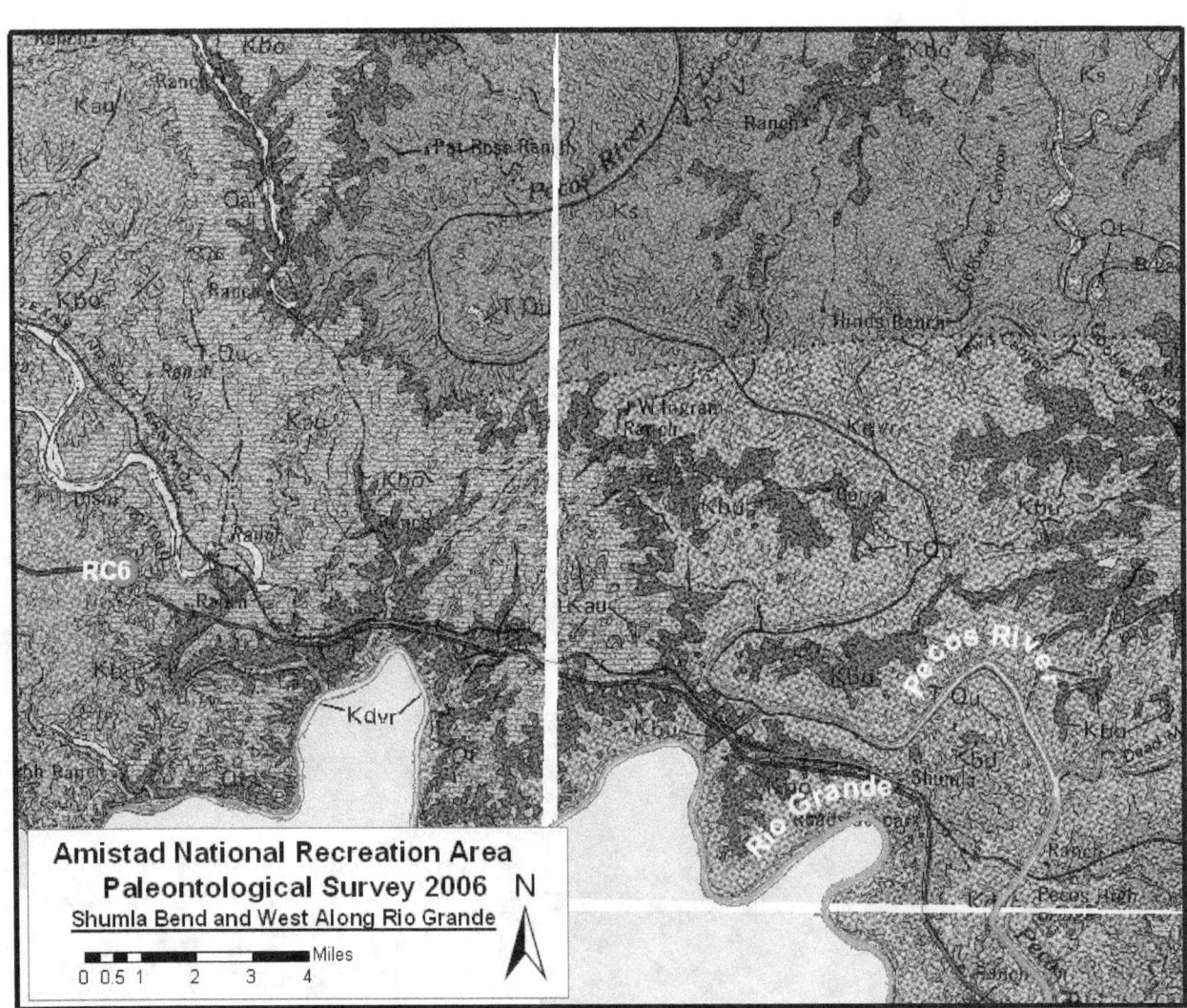

Amistad National Recreation Area
Paleontological Survey 2006   N
Shumla Bend and West Along Rio Grande
Miles
0 0.5 1   2   3   4

Appendix D: Geologic Time Scale

Notes: *Ma = Millions of years old. Bndy Age = Boundary Age.* Colors are USGS standard colors found on geologic maps. Modified from 1999 Geological Society of America Time scale (www.geosociety.org/science/timescale/timescl.pdf). Boundary dates and additional information from 2004 International Commission on Stratigraphy (www.stratigraphy.org/gssp.htm) and U.S. Geological Survey Fact Sheet 2007-3015 (http://pubs.usgs.gov/fs/2007/3015/).

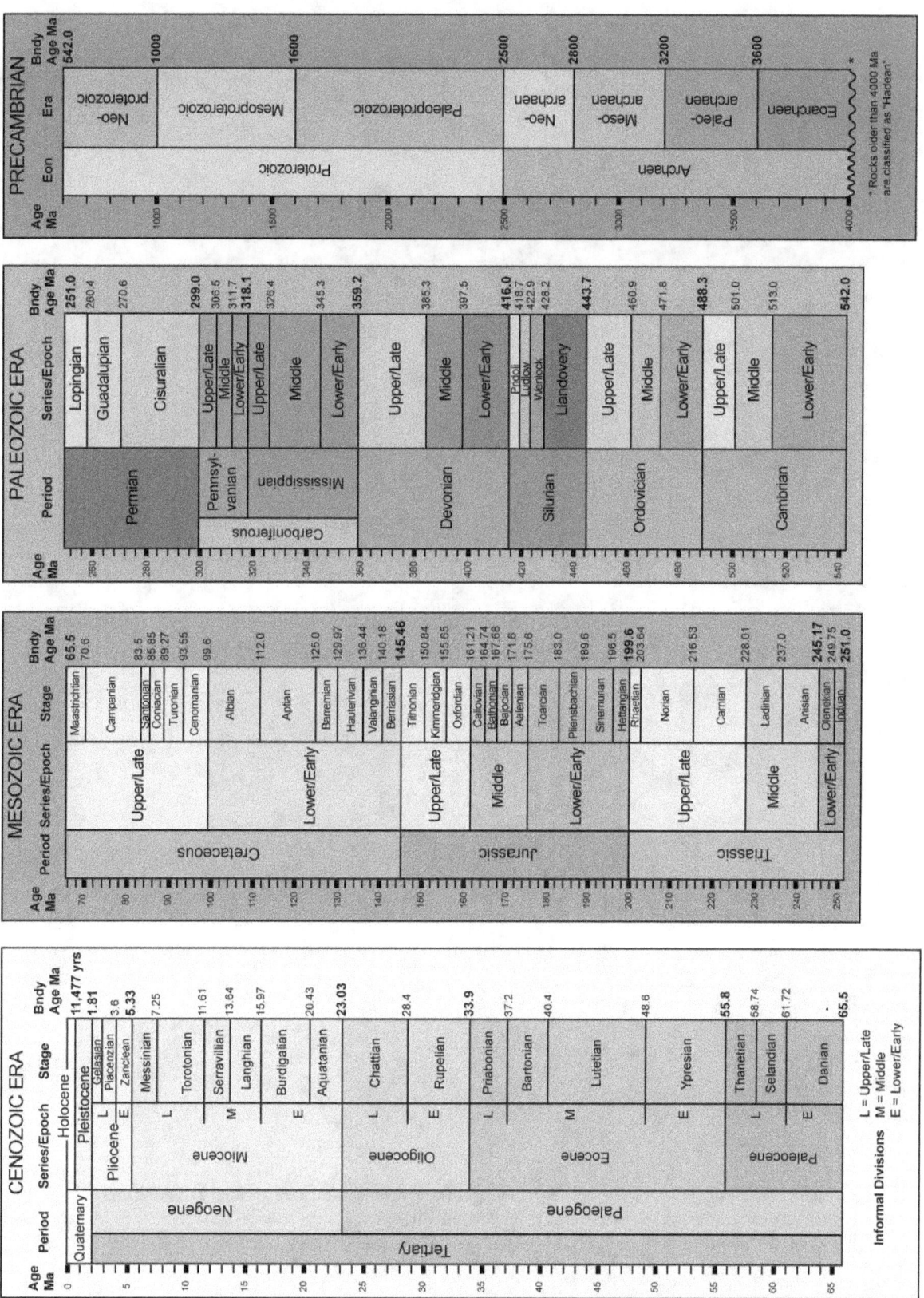

NPS 621/100191, September 2009

www.ingramcontent.com/pod-product-compliance
Lightning Source LLC
Chambersburg PA
CBHW081609170526
45166CB00009B/2897